The Universe, Life and Man

Books by P. J. Fisher

The Science of Gems

The Polio Story

The Universe, Life and Man

The Universe, Life and Man

P. J. Fisher

Illustrations by
G. Hartfield, Ltd.

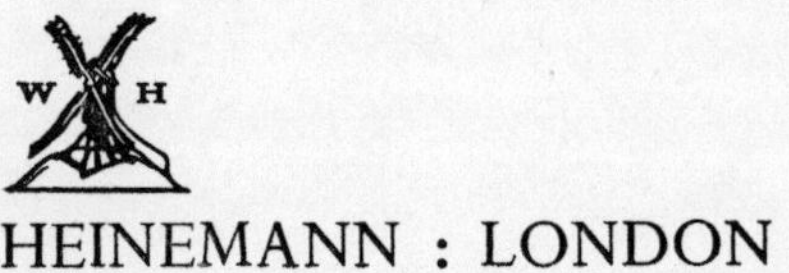

HEINEMANN : LONDON

William Heinemann Ltd
LONDON MELBOURNE TORONTO
JOHANNESBURG AUCKLAND

First published 1970

Printed in Great Britain by
Western Printing Services Ltd, Bristol

For Frances and John Lytton,
Karen and Peter Metzner,
Karen Lynn Craighead
and
Sally Fisher

Contents

List of Illustrations

5

Acknowledgements

I WISH TO EXPRESS my most sincere thanks to a number of distinguished scientists for their generous co-operation and interest during the preparation of this book. I have been able to draw on the following material published by them for data:

Mankind in the Making by Professor W. W. Howells (Secker & Warburg Ltd); *The Origin of Life* by Professor J. D. Bernal (George Weidenfeld & Nicolson Ltd); *Heredity and the Nature of Man* by Professor Theodosius Dobzhanski (George Allen & Unwin Ltd); *The Evolution of Life* by Professor Everett Olson (The New American Library Inc.)

For the privilege of including quotations long enough to require special permission grateful acknowledgment is made to the following:

Sir Bernard Lovell and the publishers The Manchester University Press for permission to quote from *Our Present Knowledge of the Universe.* Dr Ashley Montagu for permission to quote from *Man's Most Dangerous Myth and the Fallacy of Race.* The Columbia University Press for permission to quote from *Science and the Concept of Race* edited by Margaret Mead, Theodosius Dobzhanski, Ethel Tobach and Robert E. Light (Copyright © 1968 Columbia University Press). Dr Jane Van Lawick-Goodall and the publishers Baillière, Tindall & Cassell for permission to quote from *The Behaviour of Free-Living Chimpanzees in the Gombe Stream Reserve.*

I wish especially to express my gratitude to Dr Bernard Campbell, University of Cambridge, Professor Theodosius Dobzhanski, Rockefeller

University, New York, Professor John Keosian, Marine Biological Laboratory, Woods Hole, Massachusetts, Mr Patrick Moore, F.R.A.S., F.R.S.A., Professor Everett Olson, University of Chicago, Dr H. P. Palmer, University of Manchester and Professor James Scott, Queens University Belfast, for their constructive criticism of various parts of the text.

My grateful thanks also to the following for permission to reproduce or adapt illustrations:

Mr Patrick Moore, author of *Astronomy*, and the publishers, The Oldbourne Book Company Ltd. Sir Bernard Lovell, author of *Our Present Knowledge of the Universe*, and the publishers, The Manchester University Press. Dr Jane Van Lawick Goodall, author of *The Behaviour of Free-Living Chimpanzees in the Gombe Stream Reserve*, and the publishers, Baillière, Tindall & Cassell. The Houghton Mifflin Company, publishers of *The Biological Sciences Curriculum Study: Molecules to Man*. The British Museum (Natural History).

Preface

LIFE ON EARTH can now no longer be regarded as a unique phenomenon. I have, therefore, devoted Chapter 1 of this book to our solar system and the universe, 'the wider environment of life'. At first scientists believed that the ionosphere, which extends to a height of about 400 miles above the earth surface, could be regarded as the absolute outer limit of the terrestrial environment. Recent discoveries have changed all this. Within a decade space probes have penetrated deep into the universe and new evidence from space is throwing increasing light on the influences this 'wider environment of life' may exert on the earth and its cargo of living creatures.

Chapter 2 is devoted to the origin of life. Most authorities agree today that life on earth began on earth, but its origin remains one of the oldest and most elusive problems of biology and chemistry and has engaged the attention of some of the keenest minds in philosophy and science for more than 2,000 years. During that time several explanations have been attempted and, in the course of the last 200 years, many theories have come in for searching scrutiny. By now, most of them have been abandoned because they lack a scientific basis. We still do not know how life originated, but some theories appertaining to the problem have now been supported by experimental studies.

Once life had become established on our planet it became subject to many changes. These we have called the evolution of life, an experimental process which has continued until the present time. In Chapter 3, I attempt to describe this process as simply and directly as possible. Life has evolved from simple chemical beginnings to the millions of different organisms that have passed through the geological eras, and it is continuing to evolve. Each of these organisms has followed its own evolutionary path, and it would clearly be impossible to describe them all. But the processes of evolution apply equally to all organisms, and it is with these processes that I have mainly concerned myself in the third chapter.

Man's mammalian beginnings date back about 70 million years, to simple shrew-like animals. In Chapter 4, I trace the mammalian stage of our evolution to a period in time 35,000 to 40,000 years ago when a threshold was reached in brain development. After this the efficiency of culture in transmitting information from generation to generation became sufficient, and further brain development leading to increased individual intelligence conferred no particular survival advantage upon the possessor. By a gradual process man had become the thinking being he is today.

Chapter 5 is devoted to a discussion of man's racial origins and diversities. The term 'Race' as applied to man is a wide concept. It instils fear into some and generates hatred in others. I hope that my evaluation of 'Race' and its origins will help towards greater understanding of this controversial subject.

The last chapter of this book is devoted to an appraisal of the effects of human cultures on our present evolution. Of all cultures, science, more than any other, is influencing man's evolutionary path. The advancement of science is chiefly the business of specialists. As science expands, so these specialists tend increasingly to concentrate on their own narrow spheres of research, sometimes losing sight of one of the principal aims of science, a *real* improvement in the conditions of mankind. The last chapter may ask more questions than it answers.

P.J.F.

1

The Wider Environment of Life

The Universe and the Solar System

A DISTANCE of one light year—a moderate distance in cosmological terms—measures approximately 5,880,000 million miles. This is the distance light travels through space in one year. It seems so fantastic that, in earthly terms, it becomes practically meaningless. Yet the nearest star outside our solar system, *Proxima Centauri*, is 4·3 light years away from us. This means that it would take an astronaut in a rocket ship travelling at 25,000 miles an hour 115,452 years to reach the nearest star outside the solar system. It can be appreciated, therefore, that even at the present rate of technological development, it is extremely unlikely that in the foreseeable future man will be able to travel far beyond the solar system.

Although the exploration of our own planetary system will probably be accomplished within the next hundred years, the existence of life elsewhere in the universe is a calculated guess, and is likely to remain so for a very considerable time.

All we really know to date about life in the universe is that the earth, with its rather fragile load of watery living things, is favourably placed within the solar system, being the right distance from a relatively small star, the sun, to make the chemical processes associated with life possible.

The sun, one of a vast number of stars situated in our galaxy, the Milky Way, is positioned in the dusty edge of one of its outer spiralling arms. Millions of star-filled galaxies exist in the visible universe and our own planet earth is but an infinitesimal piece of matter in the vast cosmos. It is, however, the home of life as we know it and, therefore, from our point of view, is among the most important pieces of matter in the universe.

The Solar System

FOR THOUSANDS OF YEARS man has been fascinated by the star-filled heavens. Aristarchus, the Greek astronomer (about 270 B.C.) of the Aegean island of Samos, was one of the first philosophers to challenge seriously the idea that earth was the immobile centre of the universe about which the heavens daily rotated. His assertion that the earth revolved around the

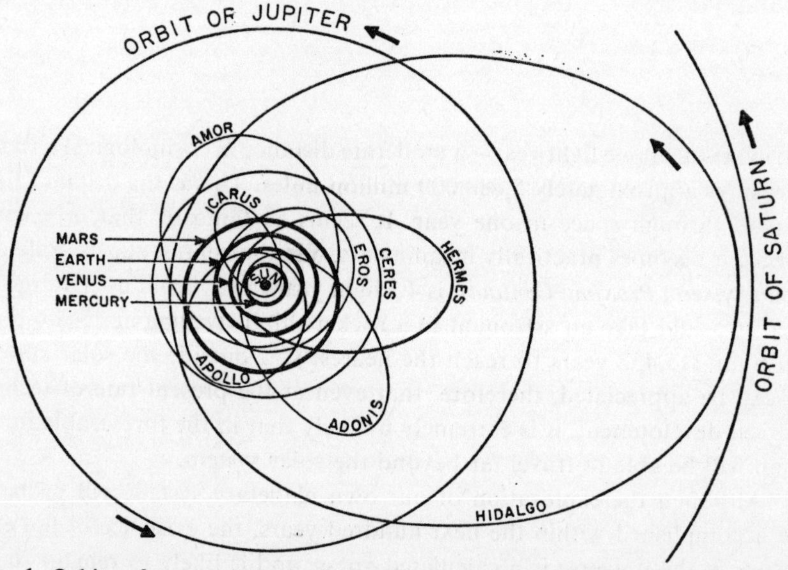

FIG. 1. Orbits of some of the major and minor planets. (After Patrick Moore.)

sun was ridiculed by his contemporaries, and it was not until much later that the Polish astronomer, Nicolaus Copernicus (1473–1543), in his famous publication *De Revolutionibus Orbium Coelestium*, finally deposed the earth from its position as the centre of motion by stating that it, and the other planets of the solar system, moved about the sun.

Copernicus's ideas, however, were still bound by traditional Greek beliefs that the motion of the heavenly bodies must be a combination of uniform circular motions.

Further advances in the understanding of our planetary systems were made possible by the superb studies of Tycho Brahe (1546–1601), one of the greatest practical astronomers of the late Renaissance, who, from his observatory on the Baltic island of Hven, studied the motions of the planet Mars. Later, as court mathematician in Prague, his observations paved the way for the German astronomer, Johannes Kepler (1571–1630), who showed that Mars, the earth, and the other planets move in ellipses about the sun.

When the observations of the Italian astronomer and mathematician, Galileo Galilei (1564–1642), revealed the satellites of Jupiter and phases of Venus, the old idea that the earth is the only body about which all others must be considered to rotate had to be abandoned.

In 1687, one of the greatest figures in the history of science, Isaac Newton (1642–1727), published the *Principia.* This monumental work explained for the first time the way in which a single mathematical law could account for phenomena of the heavens, the tides and the motion of objects on the earth. From his universal law of gravitation and laws of motion, an exact description of the entire solar system could be derived.

This consists of the sun—a medium-sized star—the planets and their satellites, together with a complex assemblage of comets, minor planets, meteoric bodies, inter-planetary dust and gas all revolving around the sun itself in closed orbits under the dominating influence of its gravitational attraction.

The sun, the dominant member of the solar system, lies in approximately the same plane as all the planets, and they each move in the same sense. The sun rotates around its axis, completing a turn every 25·4 days at the solar equator, and every 29–30 days near the poles. The planets move around the sun in elliptical orbits which depart only slightly from circles and themselves rotate in the same direction. If they have moons most of these also revolve in the same direction.

The arrangement of the planets in relation to the sun shows a certain mathematical pattern. First come the four terrestrial planets, Mercury, Venus, the Earth and Mars, in that order. Then follow the four giant planets, Jupiter, Saturn, Uranus and Neptune, and finally the outermost member of the solar system, Pluto, which does not seem to fit in with any of the others. Pluto, in fact, may not be a true planet at all, and it has been suggested that it could be an escaped moon of Neptune.

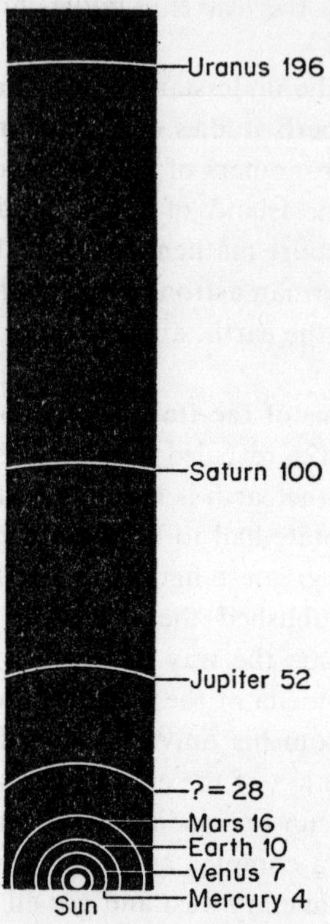

FIG. 2. Distances of the planets from the sun according to Bode's Law. The gap corresponding to No. 28 led to an organized search and the discovery of the first four minor planets. (After Patrick Moore.)

Mathematical calculations suggest there should be another planet between Mars and Jupiter. Astronomers have discovered that this space is occupied by a belt of asteroids or minor planets, which range in size from about 500 miles in diameter to less than one mile. Their total mass is estimated at only 1/3000th that of the earth. This might suggest that the missing planet once existed but disintegrated for reasons unknown.

On the cosmological scale the distances between the sun and its planets are small. Mercury, the planet nearest to the sun, is 36 million miles away from it, while 3,666 million miles separate Pluto, the farthest planet, from the sun. Our earth, the third nearest planet to the sun, is 92,868,000 miles from the sun. Thus, it takes only 8·6 minutes for sunlight to reach us on earth, and a little over six hours to reach Pluto, the outermost member of the solar system. If we reflect for a moment that some objects recently discovered in the universe may be as far as 7,000 to 8,000 million light years away from us, distances within our own solar system, by comparison, seem really quite neighbourly.

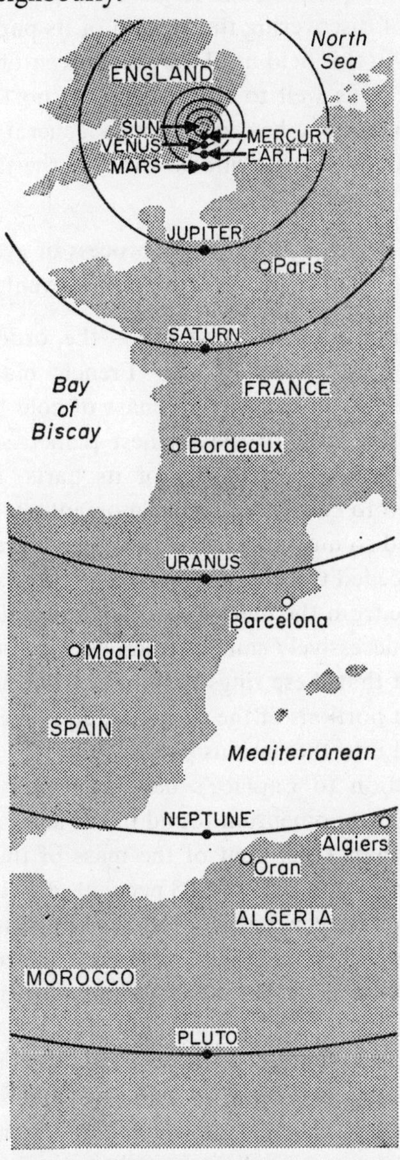

FIG. 3. Scale of the solar system. The distances of the various planets from the sun are shown on this scale drawing. (After Patrick Moore.)

The solar system, as we have mentioned, shows a remarkable degree of order, and it is now generally accepted by astronomers that it must have had its origin in a single cause. Professor Harold C. Urey has pointed out that the problem of explaining the origin of the solar system is far more difficult than that of discovering the motion of its parts, and indeed, little progress was made in this field until the present century.

At this stage it is as well to emphasize that no theory regarding the origin of the solar system has as yet won general acceptance, for all involve some improbable assumptions. Most of the theories fall into two general types:

(1) origin by an orderly process of evolution,
(2) origin by some catastrophic event.

The classic example of the first type—the orderly process—is the nebular hypothesis advanced by the French mathematician, Pierre Laplace, in 1796. He postulated a vast mass of cold gas in slow rotation extending beyond the orbit of the farthest planet. As this nebula contracted under the mutual gravitation of its parts, its rate of rotation necessarily increased to conserve angular momentum. Eventually, the rate of rotation increased so much that the centrifugal force at the periphery of the mass of gas exceeded the gravitational force there and caused a ring of material to separate from the main mass. As contraction continued, other rings broke off at successively smaller distances from the centre.

Laplace thought that these rings were not of uniform width all round, and that the densest portions of the rings gradually drew material to them, and thus condensed into the various planets.

The chief objection to Laplace's nebular hypothesis is the peculiar distribution of angular momentum found in the solar system. The planets which possess less than 1 per cent of the mass of the solar system have been able to acquire, by some means, 98 per cent of its angular momentum. Scientists cannot conceive a natural physical process during which the angular momentum of a system could have been so unequally distributed.

One way of getting a large amount of angular momentum into the planets is to put it there. This could be done by having the necessary energy forcibly injected from outside the system. This idea gave rise to the planetesimal hypothesis proposed by T. C. Chamberlin, H. Jeffreys and Sir James Jeans. Their theory is based on the assumption that the sun suffered a collision, or near collision, with a passing star. As the star approached, great tides were raised on the sun. When the conjectural star came close enough, matter could have been ejected from the sun in the form of long filaments which eventually split into separate parts and ultimately condensed into the planets we know.

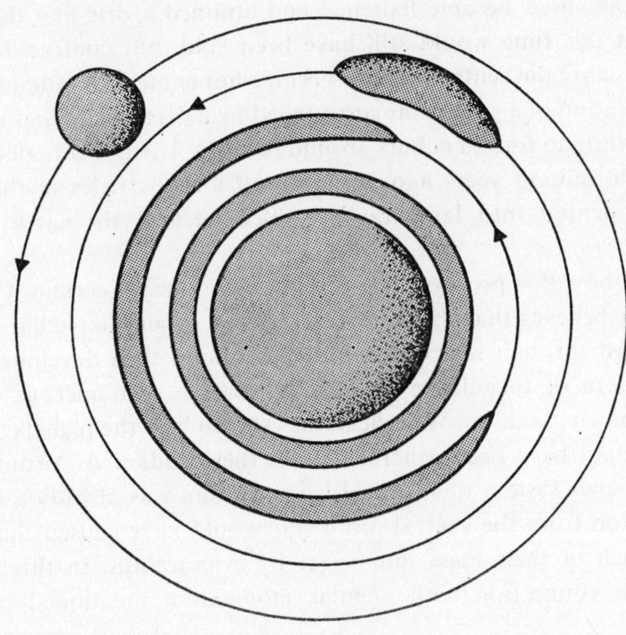

FIG. 4. Diagram to represent Laplace's hypothesis of the origin of the solar system.

Although this theory sounds fairly plausible, calculations indicate that such ejected filaments instead of revolving around the sun would probably have followed along behind the passing star to vanish in outer space. To meet this objection this particular theory was later modified, but yet another and more serious objection still remained. New evidence gathered by astronomers suggests that solar systems similar to our own are quite common in the Milky Way, although none have as yet been seen. Indeed, some authorities have estimated that in our galaxy there may be as many as a hundred million solar systems. Moreover, the collision of two stars is an extremely unlikely event owing to the tremendous distances which separate them. The nearest star to our solar system is 4·3 light years away! It is therefore highly unlikely that our own and other solar systems were formed in this way.

A more modern concept of the possible mode of origin of the solar system is that in the initial stages there existed a fast rotating cloud of cold gas and dust. Considerable turbulence would be inevitable in such a large mass, with currents forming and dying away. In due course, because of its rotation, the cloud became flattened and attained a disc-like shape. The material at this time would still have been cold, but contraction would eventually cause the central mass to become hot enough to radiate, leading to the formation of a proto-star surrounded by a disc which, in due course, would separate to form a nebula around the star. Then, in a period of time about 4,500 million years ago, the gas and dust particles in the nebula slowly condensed into large bodies which became the earth and the planets.

Exactly how this process took place is by no means certain. C. F. von Weizsäcker believes that the sun captured part of an interstellar cloud of gas and dust through which it passed. The cloud then developed an intricate pattern of turbulence with the formation of numerous elliptical currents known as eddies. Weizsäcker has shown how the planets and their satellites could have been generated from these eddies. According to his theory the solar system was formed when the sun was already a hot star. The radiation from the central proto-star would have caused the planets to lose much of their mass into space by evaporation. In this way the difficulty in connection with angular momentum mentioned earlier is avoided.

Early this century, E. E. Barnard, the American astronomer, discovered certain dark patches in front of the great diffuse nebulae that occur throughout our galaxy. These spots proved to be clouds of interstellar dust and gas which had about the mass of the sun and occupied an area of about 4½ light years, the approximate distance between the sun and the nearest star.

Lyman Spitzer, Jr, suggested that if such large masses of interstellar dust and gas exist in space, they should be pushed together by the radiation of neighbouring stars. After a period of millions of years, when the dust and gas particles have become sufficiently compressed, the force of gravitation would cause the whole mass to collapse and the pressure and temperature thereby created in its interior should be high enough to start a thermonuclear reaction—a process which is currently taking place in our sun, and provides its source of energy.

Based on these ideas, Professor H. C. Urey proposes the theory that the sun came into existence simultaneously with the rest of the solar system, and that the whole system began at a low temperature. Urey believes that if a star, such as the sun, resulted from a process of this kind there might be enough material left over to make the solar system. His approach to the origin of the solar system is that of the chemist rather than the physicist and mathematician, and since to date the subject has received most consideration from scientists of the latter two disciplines, it is therefore of particular interest.

Urey supposes that vast snow-storms consisting largely of water, ammonia and hydrocarbons such as methane, must have raged over regions as great as those between the planets today. Most of this material condensed in solid or liquid form in parts of the planets which were then in the course of formation, until eventually substantial objects consisting of water, ammonia, hydrocarbons and iron were formed. Urey thinks that some of these bodies must have been as big as the moon: indeed, the moon may have originated in this manner.

The moon also may bear evidence of the existence of colossal chunks of iron in interplanetary space. The mountains, gashed by several long grooves which encircle a lunar plain called *Mare Imbrium*, seem to indicate that this strange formation was created by the fall of a very hard body of about sixty miles diameter. Suggestions as to the nature of this material have been made by Robert S. Dietz and R. B. Baldwin, who believe that the strange grooves gashed into this lunar mountain formation must have been cut by an iron nickel alloy embedded in the main body of the object. Proof that large objects of iron still float through interplanetary space is borne out by the fact that occasionally one of them crashes into the earth as a meteorite. Such meteorites almost certainly originate from the asteroid belt between Mars and Jupiter.

Most astronomers today believe that the original material from which the planets condensed was relatively cold, and the earth was not a molten mass. Professor Urey puts forward the interesting theory that the compression of gases in the contracting planets eventually generated high temperatures that melted silicates—the compounds that form much of the earth's crust. The same high temperatures in the presence of hydrogen reduced iron oxide to iron which sank through the silicate and accumulated in large pools. The solid matter thus formed probably consisted of bodies much smaller than the present planets. These small bodies of silicates and iron were eventually collected together by gravity and collided with each other to form the present planets.

When this process was completed the gaseous envelopes surrounding the planets nearest the sun—Mercury, Venus, the earth and Mars—were gradually blown into interstellar space by radiation pressure from sunlight, since the lighter gases could not be retained by the low gravity of these planets.

For a while, until eventually they were left in the form of solid rocky bodies with negligibly small layers of atmosphere, these planets may well have looked like the comets of today, with giant luminous tails trailing behind them. (The atmosphere of the earth and inner planets is likely to have been of much later origin, and brought about by the liberation of gases from the solid crust.) The giant planets, Jupiter and Saturn, however, retained their gases because of their greater gravitational attraction; they even retained the exceedingly volatile hydrogen and helium gases. Finally, the planets Uranus and Neptune lost much of their hydrogen, helium, methane and neon, and retained only the less volatile substances such as water and ammonia.

These theories check with the present densities of the planets. Supporting evidence that some of the planets at least could never have been in a molten state comes from Mars, which, according to observations and calculations, contains about 30 per cent iron and nickel, both substances being distributed more or less uniformly throughout its mass. Had Mars been molten at one stage of its history, the heavier elements, iron and nickel, would have gravitated towards its centre.

The Birth of a Star

THE YEAR 1936 saw a new star born in a dust cloud in the constellation of Orion. It has since been named *FU Orionis*. Such events are so rare that they are not likely to be observed from earth more than once or twice every five hundred to a thousand years.

A. G. W. Cameron first recognized the rapid nature of the collapse stage of an interstellar dust cloud which might finally culminate in the ignition of a star. His theory has since been supported by more detailed studies of Chushiro Hayashi and T. Nakano of Japan.

If we can imagine for a moment an interstellar cloud of dust and gas, with about the mass of our sun, condensing by the force of starlight into a star occupying a space equal to that of our solar system, a stage will be reached eventually when gravitational attraction takes over. At this stage it will still be a comparatively cool and dark cloud, but as it shrinks further some of the energy thus released will no longer serve to heat the gas cloud but instead will carry out such internal work as breaking up hydrogen molecules and ionizing atoms. This diversion of energy will result in a reduction of gas pressure inside the dust and gas cloud until it can no longer support the outer layers of the cloud and the whole mass will rapidly collapse. Within a period of less than six months it now shrinks from the size of the solar system to a ball about a hundred times the size of the sun. The collapse is finally halted by a build-up of heat and pressure within the cloud, at which point a new star has been born with a luminosity of about a hundred times that of our sun.

E. E. Beckling and G. Neugebauer recently discovered a dark cloud in the Orion nebula which emits infra-red radiation. Their exciting idea is that this object may represent a still relatively cool cloud of stellar gas in the pre-collapse stage. Hayashi and Nakano have calculated that a cloud of this size and mass is likely to complete transition from a large cool body to a visible star within about twenty years. It will thus be of the greatest interest to scientists to find whether their predictions prove correct and a new star like *FU Orionis* is born in the next decade or two at the site of the Beckling-Neugebauer dust and gas cloud.

Our Sun

THE EVENTS described may have been similar to those which led to the formation of our sun. If so, it would have taken another 50 million years before the contraction stage was complete, and the sun settled down to the steady hydrogen-burning star we know it is today.

Life on earth could not exist unless the sun continues to shine, but it is equally important that the sun gives out its heat at a very steady rate. The term used for the steadiness of the sun's supply of energy to the earth is known as the *solar constant*. This constant can be determined experimentally by measuring the amount of heat energy which falls on the earth surface per square centimetre per minute, allowing a correction for the absorption by the earth atmosphere. The actual figure has been worked out to be nearly two calories per square centimetre per minute. This figure could not have changed to any appreciable extent in the last 2,000 to

3,000 million years—and this, from the evidence of radioactive measurements on fossil-bearing rocks, is the approximate duration life is said to have existed on our planet.

A single-celled fossil alga found in a Canadian Gunflint formation has now been dated to be about 2,000 million years old. Had the solar constant doubled or halved in this immense span of time, life on earth almost certainly would have ceased to exist. It seems, therefore, logical to believe that both the sun's distance from the earth (93 million miles) and its energy output in the form of radiation have changed remarkably little in the last 2,000 million years.

It has been calculated that the sun's energy output per year is about 3×10^{33} calories. This means that each 2 grammes of the sun's mass (2×10^{27} tons) generate 3 calories per year. In itself, this figure does not seem so remarkable until we remember that this process has been going on a good deal longer than 2,000 million years. In fact it means that each 2 grammes of the sun must have generated in excess of 6,000 million calories at a more or less steady rate. This enormous energy output could not have been sustained for any length of time by the ordinary forms of combustion which are familiar to us, considering that 10 grammes of an ideal coal and oxygen mixture would produce only about 20,000 calories, if completely burned. Thus, if the whole mass of the sun had consisted of such a mixture it would have been burned up completely in a little over 1,900 years. Obviously, therefore, the explanation of the sun's source of energy must be found elsewhere.

The energy problem of the sun has puzzled scientists since the latter part of the nineteenth century. The German scientist H. L. F. Helmholtz (1821–94) advanced the theory that the sun's gravitational contraction over a long period of time was responsible for the source of its energy. This process could not have taken longer than about 20 million years, and although Helmholtz thought this to be the approximate age of the sun, geologists rejected this period as totally inadequate.

The development of nuclear physics during this century has at last solved the mystery of this interesting problem. Scientists realized about twenty years ago that it was the thermonuclear process whereby hydrogen is converted into helium which was the main source of the sun's energy. Eight hundred million tons of hydrogen are thus converted into helium in the solar furnace every second, with an accompanying loss of mass of some 6 million tons per second.

The transformation of only 1 per cent of the sun's mass from hydrogen into helium supplies enough energy to keep it shining for 1,000 million years, and at its present rate of hydrogen burning the sun could conceivably continue to shine for another 50,000 million years. Against this, however, is the fact that the sun cannot consume more than a fraction of its hydrogen supply in this manner, and thus the process is not likely to continue for more than about another 8,000 million years.

The sun's chemical composition has been of particular interest to scientists, but its investigation had to wait until the science of astronomical spectroscopy was developed. In 1666, when Isaac Newton observed the separation of the colours of the rainbow in a prism, a beginning was made. Real progress in astronomical spectroscopy, however, became possible when in 1814 Joseph von Fraunhofer used a telescope in conjunction with a prism and a distant slit. With his instrument, Fraunhofer observed and mapped 754 dark lines in the spectrum of the sun.

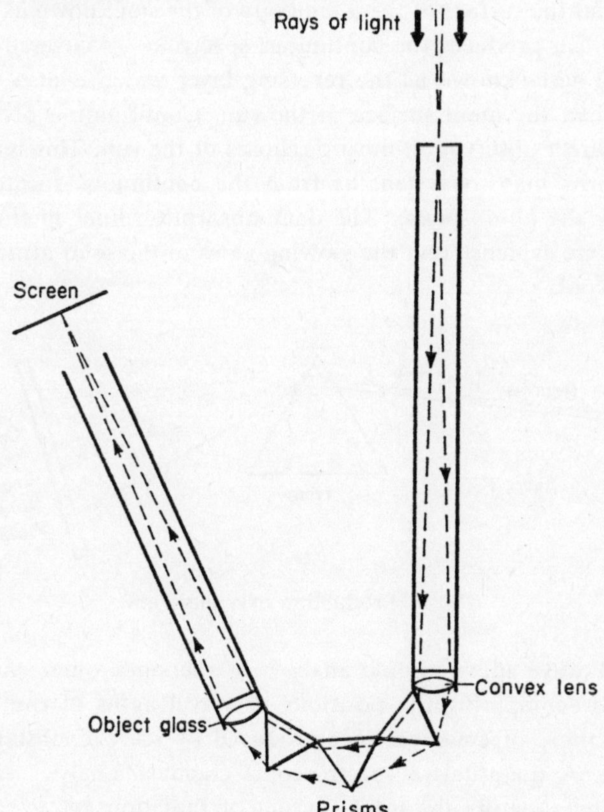

FIG. 5. The principle of the spectroscope.

Since those early days the spectroscope has become the most important of the auxiliary instruments used with the optical telescope. In investigations of the sun's composition the spectroscope can select minute portions of the flood of solar radiation for detailed study. The human eye is sensitive to only a small band of the vast range of wavelength poured out in the total radiation of the sun, but over this restricted range the word 'wavelength' is analogous to a precise specification of colour.

The sun's spectrum consists mainly of a continuous bright background on which dark lines appear. This is called an absorption spectrum. Roughly 20,000 absorption lines have been observed in the sun's spectrum and from this spectacular number more than half have been identified. Normally, the spectrum of an incandescent gas is an emission spectrum consisting of bright lines which depend on the chemical composition of the gas. The fact that the sun has such a complicated absorption spectrum suggests that the surface of the main body of the sun known as the photosphere—which produces the continuous spectrum—is covered by a layer of glowing gases known as the reversing layer which is at a lower temperature than the main surface of the sun. Confirmation of this fact is made by direct observation during eclipses of the sun. This reversing gas layer absorbs many wavelengths from the continuous spectrum below emitted by the photosphere. The dark absorption lines in the spectrum of the sun are evidence that the glowing gases of the solar atmosphere are relatively cool.

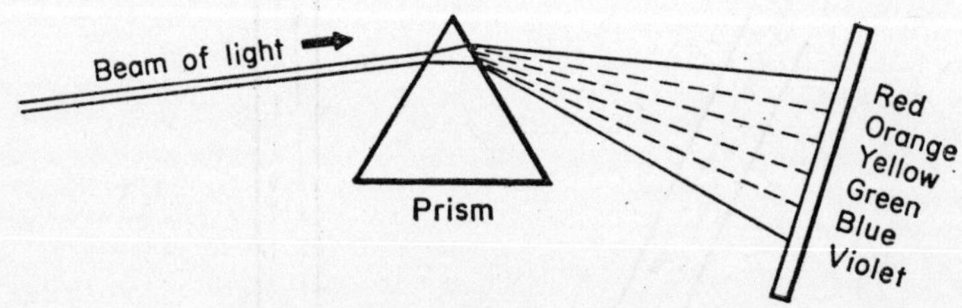

FIG. 6. Production of a spectrum.

A qualitative spectroscopic analysis of the sun's outer envelope can be made by comparing the positions or wavelengths of the absorption lines with those of emission lines produced by known substances in the laboratory. A quantitative spectroscopic chemical analysis can also be made by investigating the total amount of radiation extracted from the continuous background by the dark lines. The actual darkness of an absorption line is related in a complex way to the amount of the material that produces it. These means enabled scientists to discover that the chemical constitution of the sun closely resembles that of the earth. So far, 67 chemical elements and 18 chemical compounds known on earth have been discovered in the sun by spectroscopic analysis, but the vast bulk of the sun is made up of hydrogen and helium, and these two elements account for nearly its whole mass.

I have dwelt on the spectroscopic analysis of the sun at some length because the spectroscope is one of the most important tools of the astronomer, and astronomical spectroscopy can be applied to all stellar bodies visible through telescopes.

FIG. 7. Fraunhofer lines. These dark absorption lines appear in the spectrum of the sun. They were named after the German optician who first studied them.

Spectroscopic investigations of the visible universe have shown that all matter seems to be made up of those elements which we have found on earth. Scientists believe that the primeval material of the universe is hydrogen, the simplest of all the elements, and still the most common in the universe. Hydrogen consists simply of a proton and an electron. The more complex and heavier elements were probably formed by thermonuclear reactions in the interior of stars. Thus the basic material of our earth, both living and non-living matter, in all likelihood originated in the hot interior of some star thousands of millions of years ago.

The Earth's Environment

RECENTLY earth satellites and space probes have revealed much of the immediate environment of the earth, resulting in some startling discoveries in this scientific field since World War II. Broadly speaking, scientists believed, prior to current space experiments, that the planets and the earth were almost isolated members of the solar system moving under the gravitational field of the sun. They were isolated from each other by almost empty interplanetary space. It was, of course, well known over many years that the earth is surrounded by an atmosphere consisting chiefly of nitrogen, oxygen and some other gases all of which are held in a region of up to 50 or 60 miles altitude by the gravitational attraction of the earth itself.

When, in 1901, Marconi first communicated by radio across the Atlantic, it was obvious to scientists that some mechanism was at work to enable radio waves, which are propagated in straight lines, to overcome the curvature of the earth. If this were not so, radio communications between two points which did not have mutual optical visibility would have been impossible. Later investigations revealed that there were two main regions of ionized material consisting mainly of electrons which prevented the radio waves from escaping into outer space. This region of electrons, known as the ionosphere, acts as a reflector for radio waves over a certain range of wavelength. The radio window is limited at the long wave end by the ionosphere. Radio waves of more than a few metres wavelength suffer absorption or complete reflection in this region. This, of course, applies to both radio waves which come from outer space and those that originate from the earth.

In the years following the discovery of the ionosphere many investigations were carried out in this region, and it was found that its electron density was under close control of the sun, reaching a maximum at local midday, and being intimately connected with the sunspot cycle.

At first, scientists believed that this ionosphere, which extends to a height of about 400 miles above the earth's surface, could be regarded as the absolute outer limit of the terrestrial environment. Surprises were in store for them when the first American earth satellite Explorer I began to circle the earth. It carried a simple piece of apparatus called a Geiger counter, a device used in detecting radiation from radioactive material. This Geiger counter in the satellite formed a research project of Professor Van Allen of Iowa. He planned to study the cosmic ray particles from space before they reach the upper regions of the earth atmosphere. The Explorer I satellite had been placed in an orbit which reached well above the ionosphere. At first, the Geiger counter functioned normally, but when the satellite reached a certain altitude the Geiger counter stopped working. It began to work again when the satellite was in a different and lower part of its orbit. Van Allen then realized that the reason for the apparent failure of the Geiger counter was the extraordinary fact that in certain parts of

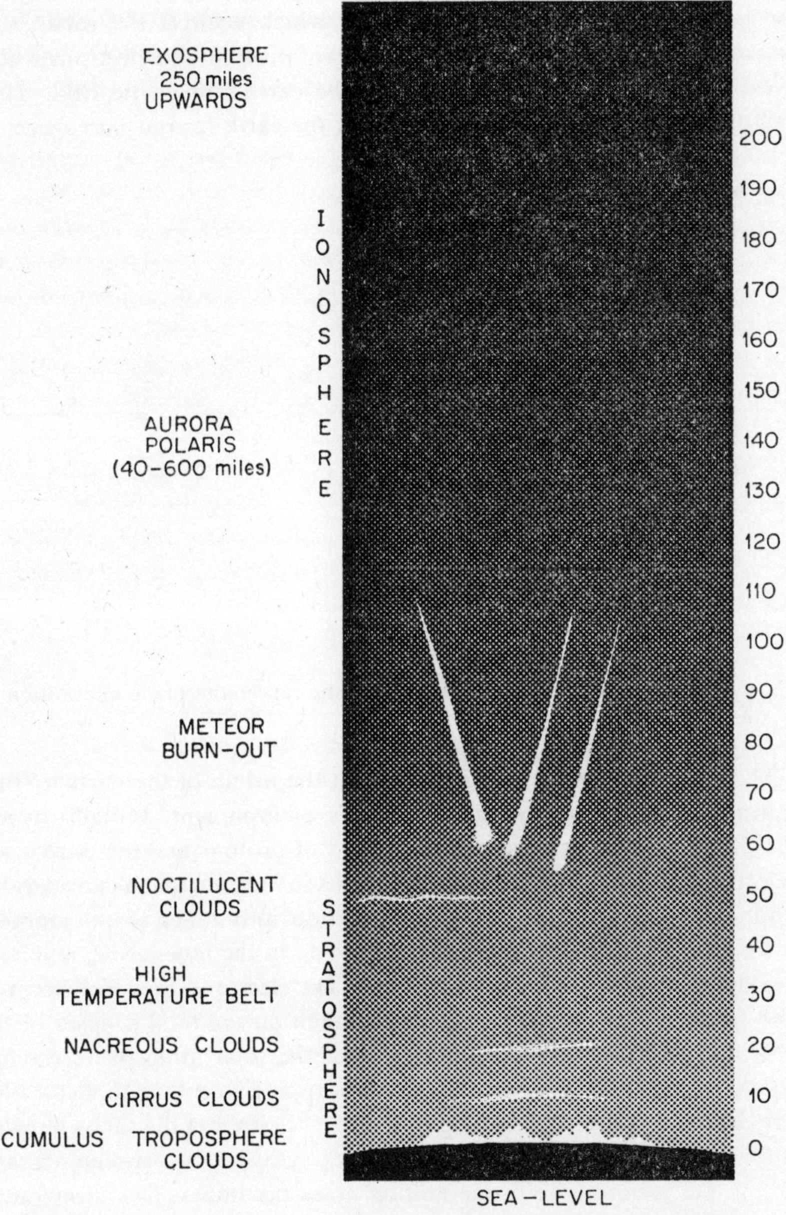

FIG. 8. Cross-section of the earth's atmosphere. (After Patrick Moore.)

the orbit the satellite was encountering an extremely dense environment of ionized material which was actually saturating the counter. In this way the now famous Van Allen radiation zones which encircle the earth were discovered. These zones are formed mostly of protons and electrons which have been trapped and accelerated by the earth's magnetic field. They reach from the immediate environment of the earth far out into space.

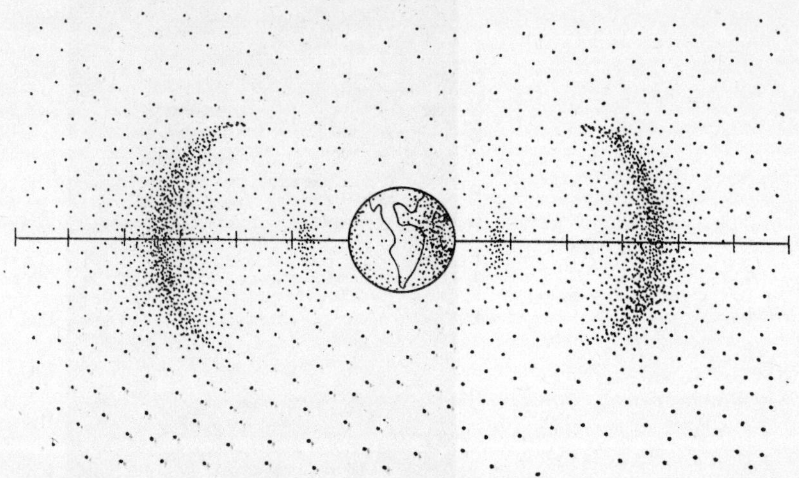

FIG. 9. The Van Allen radiation belts. The line represents plane of rotation of the earth.

There has been much discussion about the origin of the particles that are trapped in this region of the earth's environment. Initially it was believed that the inner zone was composed of protons and the outer zone of electrons. But more recent evidence seems to indicate that a considerable amount of mixing occurs and the separation into zones is not marked. On the whole, however, protons predominate in the inner zone, and so it seems likely that they have a local origin. The electrons, however, seem to come from the sun. From photographs taken during total eclipses of the sun it has been known for some time that the solar atmosphere extends into space, but only recently has it been appreciated that this so-called solar corona is constantly spilling out into space. In fact the sun is blowing out from the corona streams of electrons at velocities of between 350 and 500 km. per second (about one million miles per hour). This blow-out is known as the solar wind. The solar wind permeates the solar system, and it is now realized to be a crucial factor in controlling many of the conditions affecting the earth, the planets and interplanetary space. Indeed, it appears that the earth is enveloped in the material of the sun blown out by the solar wind.

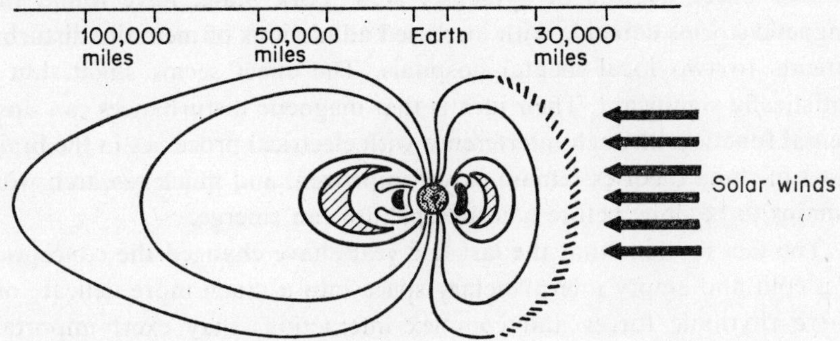

FIG. 10. Diagram to represent the zones in which electrons and protons are trapped in the earth's magnetic field. The ionosphere is represented by a thickened line around the earth. An inner zone shown blocked contains protons while electrons are predominant in the outer zone, reaching a maximum concentration in the region shown hatched. On the sunward side the region shown in heavy hatching is the magnetopause where the interaction between the earth's field and the solar wind is resolved over a small distance. On the dark side of the earth this region is ill-defined, which means that the earth has a long magnetic 'tail'. (After Sir Bernard Lovell.)

Recently the British meteorologist E. N. Lawrence has advanced a theory which depends on the influence of solar radiation on the ozone belt in the upper atmosphere. (Ozone is a gas consisting of three oxygen atoms linked together.) Lawrence claims that variations in the solar wind could influence the westerly airflow in the high latitudes. He has also observed that bad London smogs usually occurred when there were a minimum of sunspots on the surface of the sun.

The American scientists Dr G. W. Brier and Dr D. A. Bradley have detected a slight tendency for increased rainfall in the United States following the phases of the moon. Here again the problem is to think of a possible explanation for this phenomenon. Does the moon in any way disturb the solar wind? If so, how can it affect our weather?

Recently, it has been suggested by Dr Richard Head, a Boston physicist, that not only does the sun affect the planets, but the planets may affect the sun by their gravitational pull on the sun. Dr Head discovered that this would vary in a cyclic way related to the eleven-year sunspot cycle.

He also claims that certain combinations of planetary positions can give the sun a gravitational 'jerk' which may trigger a solar flare.

Interesting work is also proceeding in the medical field where there is some tenuous evidence that mental illness can be affected by solar conditions. Three doctors in Syracuse, New York State, have found that magnetic storms coincide with increased admissions of mentally disturbed patients to two local mental hospitals. The effect seems slight, but is statistically significant. Their idea is that magnetic disturbances can upset mental function through interference with electrical processes in the brain. Most of these theories remain as yet unproven, and much research work remains to be done before a clearer picture can emerge.

The fact remains that the last few years have changed the conception of a cold and empty interplanetary space into a much more delicate one where rhythmic forces and complex interactions may exert important influences on the earth and its load of living matter.

The Age of the Earth

BEFORE the present century, geologists calculated length-periods in the earth's history on the basis of the rate of deposition of sediments, the thickness of observed rock formations and similar data. Early estimates of the age of the earth ranged from 100 to 500 million years. The discovery of radioactive decay at the beginning of this century provided an entirely new and much more reliable method for dating events in the earth's history.

Radioactivity consists in the transformation of one species of atomic nucleus into another. This transformation occurs spontaneously in the case of naturally radioactive substances, and no chemical or physical condition of the environment has any effect on the rate of radioactive decay. In any given interval of time, a fixed fraction of the atoms of the mother substance decays into atoms of the daughter substance. Radioactive decay, therefore furnishes a natural and trustworthy clock with which to probe the distant past.

Lord Rutherford and his team at Cambridge demonstrated that the radioactive elements uranium and thorium ultimately decay into helium and lead. The U.S. scientist Bertram Boltwood, who worked with them, suggested that it might be possible to date the age of rocks containing uranium and thorium, providing it was feasible to measure accurately the amounts of uranium and thorium left in the minerals, and also the amounts of lead and helium present.

The processes employed in these measurements are extremely complex and leave many difficulties to be overcome. Helium is a gas and some of it might well have escaped from the rocks before they were examined. This would lead to a too low age estimate for the rock under test. Again, if the minerals contained primeval lead that was not a product of decay, but was present when the mineral was formed, a too high age estimate would result. The radioactive decay processes used in the dating of minerals take place over immense spans of time. The years indicated below represent the half-lives of naturally occurring radioactive elements used in the dating process. This means that they give the time in which exactly half the nuclei of the mother substance present at the beginning have been transformed into nuclei of the daughter substance.

Uranium 238 to lead 206 and helium	4,500 million years
Thorium 232 to lead 208 and helium	13,900 million years
Potassium 40 to argon 40 and calcium 40	1,300 million years
Rubidium 87 to strontium 87	50,000 million years
Uranium 235 to lead 207 and helium	710 million years

The above are also called primary radio isotopes because they seem to have been made at the same time as the elements first formed in our part of the universe.

Most of the precise age determinations have been made on rocks whose lead content is almost entirely derived from radioactive decay. In some cases, accurate corrections can be made for non-radiogenic lead because the lead and the uranium were sharply separated from each other when the rocks originally crystallized.

Many age determinations have been carried out, some with great accuracy. One involved an ancient monazite rock from Rhodesia whose age has been calculated around 2,700 million years. Other rock samples have given ages approximating 3,200 million years. This, then, can be regarded as the minimum age of the earth. But what about evidence for a maximum age? If it is supposed that all lead 207 has come from the decay of uranium 235, the time required to produce the present concentration of lead 207 entirely by uranium decay is about 5,600 million years. Actually, as some lead 207 was more likely formed when the elements came into being, the figure of 5,600 million years for the age of the earth is probably an over-estimate, but we can agree to a figure somewhere between 3,200 and 5,600 million years.

Much evidence about the age of the solar system has been derived from the study of meteorites and an examination of their lead content—a notoriously difficult task owing to the minute quantities present. Indeed, the only lead that could at first be detected was that present in the chemical reagents used by the scientists conducting the tests. It required the construction of a completely lead-free laboratory at the California Institute of Technology to achieve success in isolating lead from a meteorite. The technical skill employed in this work was of the highest standard, for the amount of lead present in the sample of meteorite proved to be only one part in 3 million. Results of this work gave an age for the meteorite of 4,500 million years. Similar studies seemed to confirm this age, and if it is assumed that the meteorites came into being at the same time as the earth, this provides an acceptable figure for the age of our planet.

If 4,500 million years is agreed as the earth's age, how old are the elements which make up its total mass and that of at least one part of the universe? Scientists have calculated that the elements must have a finite age not very much longer than the half-lives of the radioactive elements in existence today. Based on the present relative abundances of uranium and lead, the elements in our solar system and probably also in our part of the universe cannot be much older than 6,000 million years. It must be remembered, however, that although the main process of element formation probably took place by thermonuclear processes in the interior of stars during the early stages of the evolution of the universe, further adjustments have undoubtedly occurred over subsequent periods. For example, spectroscopic data would seem to indicate that so-called 'population 1' stars in the spiral arms of many galaxies have a much larger concentration of heavier metallic elements in their atmosphere than the stars of 'population 2', found in the central regions of the spiral galaxies. Why such differences exist nobody knows, but it is possible that evolution of the elements, like that of the universe, was not an instantaneous process but stretched over an immense span of time.

The Universe and the Galaxies

THE BUILDING of new and better optical telescopes during the last forty years has added considerably to the knowledge of stars and galaxies. Less than half a century ago it was popularly believed that the solar system was the centre of the universe. Today it is accepted by all astronomers that the sun is a typical average member of a large assembly of stars in the Milky Way.

Harlow Shapley, the American astronomer, was able to show that the sun and the solar system could not be at the centre of the system of stars in the Milky Way, and today we know that the stars in the Milky Way are not distributed symmetrically about the sun but are arranged in a flattened disc with the sun far from the central regions. The galaxy contains some hundred thousand million stars, of which the sun is a typical member. The dimensions of the Milky Way are such that starlight takes about one hundred thousand years to traverse the disc. The dimension through the disc varies from about twenty thousand light years near the central region to about one thousand light years or so at the extremities. The stars are not arranged uniformly in the disc but form a series of spiral arms which radiate from the galactic nucleus. The entire system is rotating so that at the distance of the sun from the centre—about thirty thousand light years—the system makes a celestial revolution once every 225 million years.

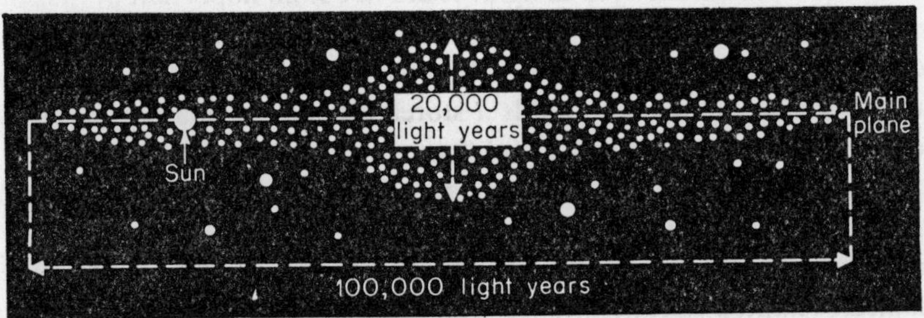

FIG. 11. Diagram of our galaxy as it would be seen edge-on. The galactic centre lies about twenty-five thousand light years away from us. (After Patrick Moore.)

Shapley's work was soon followed by observations of another American astronomer, Edwin Hubble, with the Mount Wilson telescope, then the world's largest telescope, with a reflecting mirror of 100 inches in diameter.

Hubble discovered that certain objects which had been believed to be nebulous regions of diffuse gas in the Milky Way were, in fact, star systems (galaxies) in their own right situated in space far beyond the limits of the Milky Way. The best known of these star systems, the spiral galaxy in Andromeda, is so far distant that light from it has taken 2 million years on its journey towards us. Hubble showed that the Milky Way system, with its hundred thousand million stars, was but a small part of the universe open to exploration with powerful new optical telescopes. Photographs taken at Mount Palomar with the even larger telescope containing a 200-inch reflecting mirror revealed a universe populated by as many as one trillion galaxies (1 followed by 18 noughts), reaching several thousand million light years into space.

Until very recently our total knowledge of the Milky Way and the remote parts of the universe beyond it has depended upon the sensitivity of our eyes, which can visually identify only a very small part of the electromagnetic spectrum which we call light. This is the optic region of the electromagnetic spectrum extending from the red to the blue which is so apparent in an ordinary rainbow. Radiation of wavelengths longer than red light (infra-red and wireless waves), or shorter than blue light (ultraviolet and X-rays, is invisible to our eyes and is partly absorbed by the earth's environment.

According to the fundamental laws of physics, hot bodies such as stars pour out most of their energy in the narrow part of the spectrum which we can recognize with our eyes as light. It did not seem important at first that we were unable to study the radiation in other parts of the spectrum. Only later was it realized that the invisible radiation that traverses the universe in the form of radio waves holds a storehouse of information which can help to resolve many complex problems.

It has been known for some time that hot bodies such as the sun emit radio waves in addition to the light energy they pour out in the visible part of the spectrum. These radio-wave emissions might well have remained of limited astronomical interest, but for some remarkable observations made by the American engineer Karl Jansky in 1930–31. While working with the Bell Telephone Laboratories, Jansky was asked to investigate possible sources of interference with long-distance radio communications. The remarkable discovery he made was that even when there were no obvious sources of interference, such as thunderstorms, there remained a noise signal in his receiving equipment. Moreover, he found that the strength of this signal varied throughout the day and did not recur precisely within a twenty-four hour period, but one of 23 hours 56 minutes. This is known as a sidereal day and represents the period of rotation of the earth with respect to the stars.

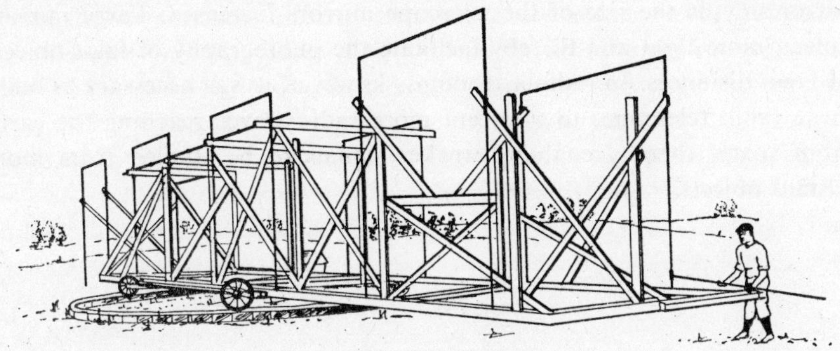

FIG. 12. Karl Jansky's aerial, the first true radio telescope. (After Patrick Moore.)

From this simple observation he rightly concluded that this noise signal, with a period of the sidereal day, must be resulting from the reception of radio waves originating outside the solar system amongst the stars and from the stars. Eventually Jansky realized that there must be something most unusual about these radio waves from outer space, because his equipment was not sufficiently sensitive to detect radio waves emanating from well-known hot stellar bodies in space. His conclusion was that either the stars themselves, or the interstellar gas between them, must have some peculiar properties that enabled them to emit powerful radio waves of much greater strength than would normally be emitted by hot bodies at the temperatures known to exist in the stars.

For a number of years little further work was done on this intriguing problem. Then the American amateur investigator Grote Reber built himself the first of the now familiar bowl-shaped radio telescopes. When he directed this telescope at the individual bright stars in the sky (and this included the sun), he found no increase in the strength of the radio emission. When it seemed that radio signals came from areas between the stars, Reber concluded that the radio waves originated in the ionized hydrogen gas which fills the space between the stars.

This, then, was the state of affairs that existed when the presence of these cosmic waves was first brought to the attention of scientists during the last war.

It was soon realized that the technical problems facing astronomers working with radio-wave bands—wavelengths a million times longer than the wavelength of visible light—were strictly analogous to those with which the optical astronomers had to contend. Progress was made in optical astronomy as the size of the telescope mirrors increased. Large mirrors collect more light and thereby facilitate the photography of faint objects at great distances. In radio astronomy, however, it was necessary to build large radio telescopes to intercept more radio waves reaching the earth from space, thereby enabling weaker signals to be studied from more distant objects.

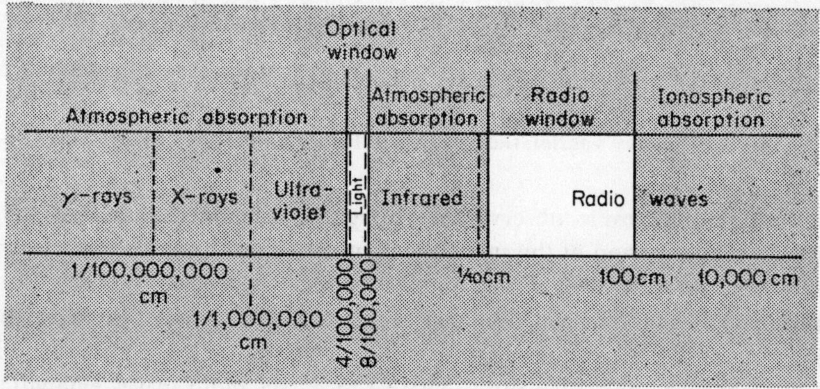

FIG. 13. The electro magnetic spectrum. The central strip is the 'optical window' indicating visible light. The earth's atmosphere will pass only visible light, a fraction of ultra-violet radiation and the radiation in the radio window. (After Patrick Moore.)

This was the basic idea which stimulated the construction of the large steerable radio telescope at Jodrell Bank, in England. This radio telescope has a reflector with a diameter of 250 feet. It is steerable and its paraboloid reflector can thus be pointed in any desired direction in the sky. Radio waves from a distant astronomical source fall on the curved surface of the reflector, thus causing the waves to be reflected. If the reflecting surface is correctly curved all the reflected rays can be made to arrive at the same focus from where they are piped through a hollow wave guide to a radio receiver. This is, of course, an over-simplification of a highly complex piece of apparatus which has helped to revolutionize our ideas about the universe.

One of the first problems to be solved by radio astronomers was to account for the numerous radio sources which seemed to be distributed throughout the sky. They were aware, of course, that remnants of exploding stars known as supernovae are powerful sources of radio waves. An example of this is the Crab Nebula, which was seen to explode in A.D. 1054 by Chinese astronomers. This expanding mass of gas is moving outward at the rate of about 70 million miles a day. The object, at a distance of about four thousand two hundred light years, is relatively close to us in the Milky Way. But there are only a small number of such supernovae remnants in the Milky Way and their radio emissions do not account for all those observed either in strength or distribution.

Many galaxies that lie far beyond our own emit radio waves with an intensity to be expected from measurements of the Milky Way emissions; but large numbers of strong radio sources which could not be accounted for by the emission of these galaxies soon became apparent.

One of the strongest radio sources in the sky lies in the constellation of Cygnus, in a region of the sky which seemed to be devoid of any outstanding visual astronomical objects. When the position of the radio source was measured with sufficient precision, it became possible to make a detailed search of the region with the 200-inch optical telescope on Mount Palomar. After an exposure of many hours, a photograph revealed a peculiar object which showed a red shift indicating that it was a distance of 700 million light years away from us. Soon, other objects of this kind were discovered. All were very faint in the optical part of the spectrum but relatively strong in their radio emissions.

Let us pause here for a moment to explain a technical term of great importance in astronomical studies—the Doppler Effect.

This term is applied to an apparent change in wavelength of light or sound caused by the motion of the source or of the observer. If a whistling train is approaching an observer, more sound waves per second enter the observer's ear than would be the case if the train were standing still; the wavelength is shortened and the whistle sounds higher pitched. When the train begins to recede fewer sound waves per second enter the ear, the wavelength is lengthened and the note of the whistle drops.

The same principle applies to light, except here the object will look too red when receding and too blue when approaching. This phenomenon is known as the Doppler Effect after the Austrian physicist who first drew attention to it in the year 1842.

Slipher first discovered that light emerging from extragalactic nebulae showed a shift of the spectral lines towards the red end of the spectrum, but it was Hubble who interpreted this red shift as a Doppler Effect resulting from the recession of the galaxies. At the limit of penetration of the optical telescopes twenty years ago galaxies could be observed 2,000 to 3,000 million light years away—and these nebulae were moving away from us at the enormous speed of over thirty thousand miles per second, which is about 20 per cent of the speed of light.

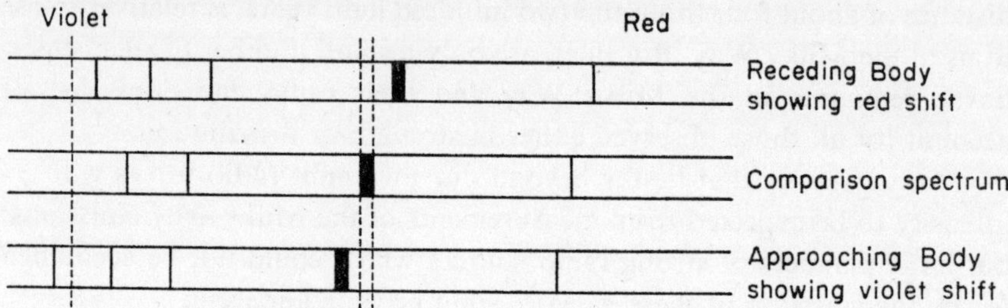

FIG. 14. The Doppler Effect. In the top drawing the light source is assumed to be receding and the spectral lines are shifted toward the red. The centre diagram is for a stationary light source drawn in for comparison; the bottom diagram, for an approaching light source, shows a shift to the violet. (After Patrick Moore.)

Such enormous recessional speeds can only mean that the universe is expanding at such a fantastic rate that by terrestrial standards it seems difficult to accept. Indeed, many people have questioned the interpretation of the red shift, but Sir Bernard Lovell has rightly pointed out that if one interprets one's observations in terms of the laws of physics as they are known to us on earth, it is not rational to abandon this principle in favour of the artificial introduction of laws which are different in remote parts of the universe.

If, then, the red shifts in the spectra of the galaxies in the remote parts of the universe really mean that these galaxies are receding from us at such fantastic speeds—and there are few astronomers who today doubt the correctness of this interpretation—we are faced with the problem of a rapidly expanding universe, and the possible origin such a universe might have.

It is now generally accepted that the basic primeval material of the universe is hydrogen, the protons and electrons in the primeval state that form hydrogen gas. Indeed, it is now established that concentrations of hydrogen gas exist in the spiral arms of the Milky Way, although the central region or nucleus of our galaxy is relatively free of the gas. This discovery has to be related to an earlier one, namely, that there are two different populations of stars in the Milky Way. Those in the central region of the galaxy are believed to be old stars, and a large percentage of the total mass of the galaxy is in a rather small region which forms this central bulge. Walter Baade called these stars in the central region population 2 stars, and the younger blue stars located in the spiral arms, population 1 stars.

The relative lack of hydrogen gas in the central region containing the population 2 stars, may be connected with the time or the process by which these stars in the central region were formed. It is noteworthy that the hydrogen gas in the central region represents only a few hundredths part of the total mass, the hydrogen gas in the spiral arms of our galaxy, where our solar system is located, accounts for nearly a fifth of the total material. It is in these regions of the galaxy, where hydrogen is still plentiful, that stars like *FU Orionis*, mentioned earlier, are now being born. In the central region, where there is little hydrogen gas, star formation seems to have ceased long ago.

Hydrogen is still the major constituent in the universe, and it would seem a natural assumption that all the material in the galaxies originated in some way from the primeval hydrogen. Only a small amount of the mass of the cosmos has so far been converted into heavy elements by thermonuclear processes in the hot interior of the stars.

If we accept the view that the universe is expanding, it follows that in a past epoch the universe must have been a much more compact entity than it is today. If we imagine time in reverse for a few hours, then it seems that we must have been many millions of miles closer to the distant galaxies than we are at present.

Since we can determine the rate of expansion by measuring the red shift in the light that reaches us from distant galaxies, it is a simple matter to pursue this idea to the ultimate state when all the galaxies and all the primeval material in the universe must have been in a densely compact condition. If we assume that there has been no change in the rate of expansion, this time turns out to be about 10,000 million years. According to this argument, the universe had a beginning about 10,000 million years ago. Unfortunately, however, we cannot be certain that this rate of expansion has remained constant throughout these great periods of time.

A solution of this type has been described as the 'big bang' theory of the origin of the universe. Here we have to imagine that about 10,000 million years ago the primeval material of the universe existed in a highly concentrated form of exceedingly high density. Then, for unknown reasons, an explosion occurred within this primeval mass which scattered it in all directions.

On this theory the expansion of the universe which we witness today would be the remaining impetus of the initial explosion. After the initial explosion, the units of the universe—whether they be clusters of galaxies or the stars which later formed into galaxies—began the process of condensation from the expanding hydrogen cloud.

Some have maintained that this initial explosion could be called the origin of time and space. Such a statement, in my view, makes no sense, because clearly human thought processes are at present quite unable to conceive a beginning to time or space. We are equally unable to imagine infinite space, yet according to all logical thought processes space must be infinite because we simply cannot conceive a limit to space in a cosmological sense.

A theory which visualizes the universe undergoing an alternate expansion and contraction in a cyclic phase has recently been further developed by Professor R. H. Dicke. This theory is based on the idea that the gravitational force may be sufficiently strong to hold and reverse the expansion of the universe.

If this is true the universe must be alternately expanding and then contracting to a very hot dense fire-ball of matter. Now we know that the behaviour of matter in atoms, crystals, stars, and the universe itself, is governed by four basic forces: strong and weak forces, essentially short range, which operate significantly only on the atomic scale, and the remaining two, the gravitational and the electromagnetic forces, which exert noticeable influences on matter of every scale. It may seem surprising that the gravitational force which is responsible for maintaining a massive body such as the moon in orbit around the earth is by far the weakest of the four interactions. It has been calculated that the strength of the gravitational force is directly proportional to the masses on which it operates. Because the amount of matter in the galaxies and in the space between them is so vast the gravitational force dominates the motions of the universe. Given enough matter in space, and the present value of the gravitational constant, there is reason to believe that the universe will only continue to expand until it reaches a certain maximum size. Thereafter, it will contract to generate a fire-ball of matter and radiation of extremely high temperatures and density, but with dimensions as small as a few light years.

Professor Dicke suggests that the argument for a state of high temperature at the start of each expansion phase is a simple one. Hydrogen is converted into heavier elements in the stellar interior by thermonuclear processes, but in its early days our galaxy is believed to have been free of the heavier elements. This implies that the temperature during the contracting phase was so high that any matter of a previous cycle was decomposed to pure hydrogen. If this is what really happened it would have required a temperature of about 100,000 million degrees centigrade.

Professor Dicke also suggests that radiation from such a high temperature fire-ball might still be detectable today as a uniformly distributed cosmic radio noise. Observations with a specially constructed radio telescope functioning at a wavelength of 3·2 cm. have given some encouraging results which support this hypothesis.

It has also been suggested that the expansion and contraction phases would take about 45,000 million years each. According to this theory we are now at the beginning of an expansion phase with the universe about 10,000 million years old. One particularly attractive feature about this theory is that a periodically expanding and contracting universe avoids the necessity for assuming a complex initial condition for its start.

Shortly after the second world war a radical new concept came into the forefront in cosmology—the theory that the universe is in a steady state and its expansion is balanced by continuous creation. It was proposed and developed by Fred Hoyle, Herman Bondi and Tom Gold. This theory had the attractive feature that the difficulty of explaining the beginning of the universe was avoided. It leads to the concept of the continuous creation of the hydrogen gas from which the galaxies are formed. Because of the expansion of the universe galaxies are constantly moving out of our field of view, and the theory proposes that new galaxies must be in the process of formation to counterbalance those that are lost to view. The force leading to the expansion of the universe in the steady state theory arises as a consequence of the creation process. The difference between the evolutionary theories of the universe, which envisage a moment of creation for the galaxies in the remote past, and the steady state theory, which involves the continuous creation of matter, is fundamental and much time has been spent determining which of these theories is the more likely one.

A possible solution to this problem lies in our ability to observe ever more distant regions of the universe. As this becomes possible the decision as to which of these theories is true may be confirmed. It is well to remind ourselves at this stage that an observation made today of a galaxy 2,000 million light years away is not a record of this galaxy as it is at this moment, but as it was 2,000 million years ago, when light and radio waves left it on their long journey through space.

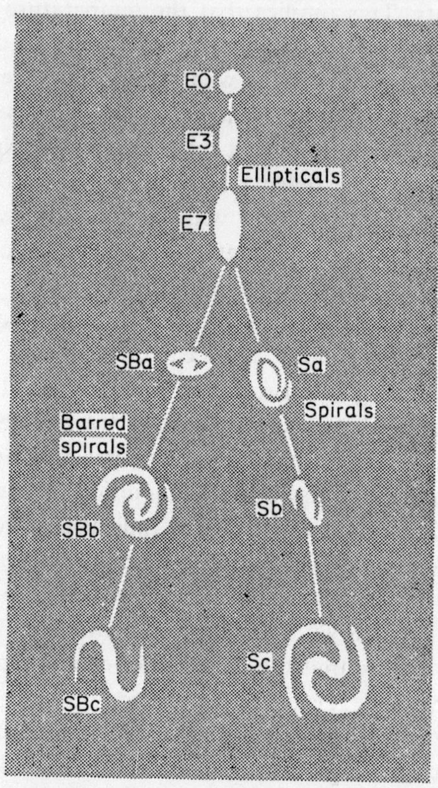

FIG. 15. Hubble's classification of the galaxies: ellipticals (E0 to E7), spirals (Sa, Sb, Sc), and barred spirals (SBa, SBb, SBc). (After Patrick Moore.)

Obviously then, because of the finite speed of light and radio waves which travel through space at 186,000 miles per second, our penetration into the depth of space is also a penetration into the remote history of the universe.

One of the problems that faced the astronomers was that their predictions about the general structure of the universe out to a distance of about 5,000 million light years differed little either in the steady state theory, or in the various evolutionary theories. Only if it became possible, therefore, to penetrate even farther into the regions of space and time might a solution to the problem of creation be found.

This farther penetration into the universe now seems to have been achieved in an entirely unexpected manner and the history of these surprising developments is closely associated with one of the research programmes conducted by Sir Bernard Lovell at Jodrell Bank. The technicalities which led to these developments are highly complex. Suffice it to say, here, that scientists thought that successful identification of distant radio sources with an optical telescope might well reveal objects farther removed in space than any so far discovered.

In 1960 a search with the 200-inch optical telescope on Mount Palomar succeeded in establishing an identification of such a radio source which had been given the Cambridge catalogue number 3 C 48. To the great surprise and astonishment of the astronomers the object visible on the photographic plate appeared to be a blue star in our Milky Way. Quite soon a number of other radio sources were identified with similar starlike objects also apparently situated in our own Milky Way. This discovery was indeed unexpected, because astronomers had thought that these radio sources originated in a remote part of the universe.

In 1962 Maarten Schmidt, at Palomar, succeeded in photographing a good spectrum of one of these blue stars, and the amazing fact was revealed that the spectrum exhibited a large red shift which indicated, as we have learned earlier, that the object was speeding away from us at a truly enormous rate.

In 1963, the results of the red shift measurement on 3 C 48 were published, and since that time the red shifts of many more of these blue objects have been measured.

Now if the red shifts of these so-called quasars are interpreted in the normal way it means that some of them must be at least 7,000 to 8,000 million light years away from us. Further, they seemed to be speeding away from us at colossal velocities approaching 80 to 90 per cent the speed of light, over 160,000 miles per second.

Particularly troublesome is the question of the sources of energy of these quasars. Even when making the most generous assumption about the efficiency of the process by which energy can be converted into light and radio waves, there is a deficiency approaching one hundred times in some of these objects, judged by any known process of energy production on earth.

The most productive energy sources on earth involve the transmutation of matter into energy in the nuclear bomb or the nuclear furnace. Assuming that the most efficient nuclear processes known to us on earth are at work in these galaxies, then the calculations indicate that the vast quantity of energy that is poured out into space from the radio galaxies and quasars requires an absurdly high proportion of the total mass-energy available.

Many recent observations have confused the outlook on cosmology and dimmed the hopes that were entertained a few years ago of a decisive advance in our understanding of the cosmos. It is possible, indeed, to construct a seemingly endless stream of paradoxes and contradictions from current observations and theories.

It is, however, now generally accepted that the radio observations of distant galaxies and quasars have undermined the steady state theory, and it will have to be abandoned in its original form. A revised form of this theory was published in 1966 by Hoyle and Narlikar.

The outcome of this present uncertainty concerning the genesis of the universe is that astronomers are generally agreed that we really do not actually know in what type of universe we are living. We can only hazard an opinion that the steady state theory in its original straightforward form cannot lead to a satisfactory solution.

If some of Professor Dicke's recent radio observations made at very high frequencies can be confirmed to originate from the primeval universe a few minutes after it exploded from the high density state, we will have some firm evidence in favour of the 'big bang' theory. The time scale which separates us from such a singular event is enormous, and may be of the order of between 7,000 and 15,000 million years.

Naturally, man will continue probing the mysteries of the universe until at least he is satisfied that all possible lines of approach to find a solution to its puzzling problems have been exhausted. Most scientists today are satisfied that we, and the rest of the living world on earth, are made of very much the same kind of stuff as are the stars and galaxies in our part of the universe. Living matter is, in fact, no more or less than a tiny portion of the universe that seems to have evolved by as yet unknown processes into the remarkable state of 'Life' which has inhabited our planet for more than 2,000 million years.

2 The Experiments Begin

The Origin of Life

EXPERIMENT IS DEFINED as 'ascertaining or establishing by trial', and indeed, it seems that life on our planet has been synthesized by trial and error from the very same elements which are found in the universe and the interior of stars.

A probe into the problems of how life might have originated raises a wide variety of scientific questions, but before we examine some of these, let us delve into the past to see what our ancestors thought about the origin of life.

Early Ideas

LIFE ON EARTH may be 3,000 million years old, and perhaps even older, though there is no direct evidence for this. It is possible that primitive forms of life were present when the earth came into existence, but only a few scientists subscribe to this hypothesis.

Long before man began to record his thoughts it is likely that he speculated about the origin of life, for man was ever curious about his environment, and when he found that a freshly killed animal was soon infested with maggots, he must have wondered from whence came those wriggling little worms. His most likely solution was that some living things evolved spontaneously, given the right conditions. For instance, maggots developed in meat. The idea of the spontaneous generation of life was no doubt conceived by the ancients, but they were wrong in supposing that animals or plants (and here I include all micro-organisms) could arise from non-living matter. Nevertheless, their ideas bore the seeds of truth. And that truth is that if life as we know it has its origins on earth, and this today is accepted by most authorities, it must indeed have arisen originally from non-living chemical substances.

Anaxagoras, the Greek philosopher (approx. 500–428 B.C.), believed that life fell from the heavens in the form of tiny seeds (*spermata*) to fecundate the earth. This idea may well have originated from observations that some plants grow from seeds. Indeed, about 9,000 years ago, such observations must have been made by the earliest known agricultural communities in the lightly wooded grasslands to the east of the Mediterranean.

The theory of *panspermia* postulates that life is universal and that the universe contains minute germs which develop on finding a favourable environment. This idea of life from outer space descending upon the earth meets with a number of serious objections. First, it explains only the appearance of life on earth but ignores the important questions of how and where life arose in the first place. Secondly, in travelling to earth from a distant planet an organism would encounter great extremes of heat, cold and deadly radiation which life as we know it could not tolerate. Meteorites, often seen at night, are particles of stone or dust that glow with heat when they encounter the friction of the upper atmosphere. They are unlikely to carry living organisms down to earth because such organisms would be destroyed before they reached the earth's surface, though the interior of large meteorites is cold on arrival.

However, the possibility that life exists on many planets in the universe which enjoy a similar environment to our own is a distinct possibility, although we have no proof to date. If there is such life, would it be able to traverse the vast spaces separating stellar systems and galaxies? This question may be answered as space probes and artificial satellites collect small quantities of debris from outer space to be analysed in the laboratories for evidence of life.

Aristotle (384–322 B.C.), whose interests extended from political theory to biology, was convinced that spontaneous generation was a certainty, and that different creatures such as insects, fish, frogs and mice could spring from suitable breeding materials like dirt and moist soil.

Aristotle's concept of the active principle seemed to offer an explanation why fertilized eggs could develop into living animals. This active principle could direct and organize a sequence of events that would produce life. Thus a duck's egg had an organizing principle that led to the development of a duckling; similarly, because of their organizing principle, frogs' eggs would develop into frogs. In *The Generation of Animals* Aristotle wrote:

> Such are the facts, everything comes into being, not only from the mating of animals but from the decay of earth . . . And among plants the matter proceeds in the same way, by spontaneous generation by natural forces; they arise from decaying earth or from certain parts of plants.

Aristotle's ideas influenced the beliefs of many generations which followed him. During the Middle Ages, people believed that geese originated from certain conifers that grew near the seashore. This belief in the goose-tree survived until the early eighteenth century. It gained credence from reports by travellers coming from the Middle East who indicated that certain trees there bore melon-like fruit which contained fully formed lambs. These extraordinary stories could well have been invented as an excuse to eat meat on fasting days—meat which was purely vegetable in origin. There are numerous drawings in existence purporting to show these meat-producing plants.

An early attack on the theory of spontaneous generation of life came from the Florentine biologist, Francesco Redi (1626–97). By controlled experiment he attempted to show that life could arise only from pre-existing life. He placed portions of meat and fish into four flasks which were then carefully sealed with gauze. A similar number of flasks were filled in the same way, but this time left open to the atmosphere. Redi's controlled experiment sought to prove that maggots could not develop spontaneously in rotting meat or fish, but that their formation was directly connected with some outside agency. Soon, he observed that although the meat in the open flasks had become infested with maggots, and that flies entered and left the vessels at will, not a single maggot was visible in the sealed containers.

Redi's experiments seemed to support the idea that life can arise only from pre-existing life. Yet although he clearly used an experimental method to disprove that maggots could develop spontaneously in rotting meat, he still accepted other claims. For example, he believed that intestinal worms arose spontaneously from decaying matter, and gall flies from plant juice.

FIG. 16. A drawing from the sixteenth century supporting spontaneous generation. 'Vegetable lambs' were produced from melon-like fruits.

In the mid-seventeenth century William Harvey, in his studies of the reproduction and development of the king's deer, made the basic discovery that every animal comes from an egg. But old ideas on spontaneous generation died hard. Even though it was proved that larger animals always came from eggs there was still speculation about the origin of such smaller ones as the infusorians. It seemed a natural conclusion that these microscopic creatures must be generated spontaneously because they seemed to appear everywhere. The perfection of the first microscope by the Dutch naturalist and microscopist Anton Leeuwenhoek (1632–1723) did much to revive some of the old ideas on spontaneous generation. Through the lens of this simple instrument scholars observed swarms of tiny organisms whose existence nobody had hitherto suspected. Although this view was not shared by Leeuwenhoek himself, others were convinced that these creatures had arisen spontaneously in the substances under examination.

In 1749 yet more 'scientific' evidence was produced by J. T. Needham (1713–81) to support the argument for spontaneous generation. In conjunction with G. L. Buffon (1707–88) he experimented with a phial filled with broth, which after being boiled was closed with mastic. The phial soon showed a lively growth of micro-organisms, convincing him that organisms can be generated spontaneously in organic substances.

L. Spallanzani (1729–99), an Italian priest and biologist of great ability, effectively answered Needham's observations by exact experiment. In 1765 he showed that if tests were carried out with the proper precautions, no growth of micro-organisms could occur. His precautions consisted in the prolonged boiling of meat broth and infusions, and in immediately sealing the vessels containing them. Mostly, his experimentations were successful and seemed to rule out the theory of spontaneous generation. Needham's experimental method had obviously been defective.

The only effective criticism made on the experiments of Spallanzani was that by heating he had not only altered the infusions themselves, but also the air contained within the phial by removing the oxygen.

To this the French chemist and biologist Louis Pasteur (1822–95) had the answer with his famous 'Swan Neck Flask' experiment.

Pasteur poured water into an ordinary flask, added to it brewer's yeast and a little sugar, then drew out the neck of the flask in a hot flame so as to give it a swan-necked curvature. Next he brought the liquid in the flask to the boil for several minutes so that steam issued freely through the open end of the neck. When the liquid in the flask was allowed to cool it was found that it remained unchanged and uncontaminated for an indefinite time. Pasteur reasoned that, when the crude air re-entered the flask during the first moments of cooling, the dust and germs were trapped in the condensed distilled water in the bottom curvature of the neck as the inflowing air bubbled through it. Dust not so removed settled in the upright portion and in the upper bend. If, after several weeks, the neck of the flask was removed by a stroke with a file, but without otherwise touching the flask, moulds and germs soon made their appearance in the liquid just as if it had been inoculated with dust from the air.

Pasteur's critical experiment demonstrated that certain mixtures, when treated in certain ways, would *not* give rise to living forms. The results of his experiment, however, did not disprove that another kind of mixture treated in some other way may produce some form of life. The negative results of Pasteur's experiment did not necessarily exclude the possibility of the success of differently designed and executed experiments. Only when all possible experimental avenues have been explored and proved futile is a negative answer valid. One thing, however, that Pasteur's experiment did show was that the original notion of spontaneous generation—that complex organisms could arise suddenly from non-living matter—was an impossibility.

When eventually the creation of living forms in the laboratory becomes possible, the experimental methods used will follow some kind of laboratory technique, and may well justify the term spontaneous generation.

Pasteur's researches discouraged many scientists of his time from further debating the origin of life lest they offend religious feeling, but one way around this problem was to assume that life had always existed in our part of the universe, and was eventually imported to our planet. The theory of *panspermia*, mentioned earlier, follows this line of thought.

Organic Matter in Meteorites

In 1864 Helmholtz (1821–94), the German philosopher and scientist, postulated that 'living matter' reached the earth by the agency of meteorites—an idea that has adherents today. It led to the repeated examination of the famous Orgueil meteorite which fell in Orgueil, France, in 1864. Chemical analysis of the Orgueil meteorite has shown that it contains 6 per cent carbon compounds. Other meteorites examined recently, some as old as 4,500 million years also seem to contain compounds of a definite organic nature. They had their origin not on earth, but on some distant planetary body.

The great similarity between compounds extracted from carbon-containing meteorites and those formed by synthetic processes from the simple gases methane and ammonia suggests that the first stages of the formation of substances connected with living processes were present everywhere in our part of the universe and are by no means confined to our planet. Subsequent stages in the development of life right up to the formation of organisms seem, however, to be confined to earth, owing to the need of all existing organisms for water and solar radiation. The controversy about the origin of organic substances found in meteorites has not as yet been resolved, and as long as it remains open it will provide the subject of much further research.

The evidence also raises the question: is the phenomenon of life and its origin part of chemical processes generally found in the universe, and thus the inevitable outcome of a form of chemical evolution?

The Biochemists Become Interested

WHEN BIOCHEMISTS first turned their interest towards those substances associated with living matter they were more preoccupied in discovering the actual chemical changes that took place within organisms than in the problems associated with the origin of life. Indeed, at the beginning of this century, these problems were still considered largely to be within the realms of the philosopher rather than the scientist.

In the early 1920s, however, a new interest in the subject was stimulated by the independent ideas of two biochemists, A. I. Oparin in the Soviet Union and J. B. S. Haldane (1892–1964) in Great Britain. Oparin's first paper on the subject, published in 1924, has remained the basis of nearly all modern theories on the origin of life. When Oparin wrote this paper, modern biology and especially molecular biology and biochemistry were in their infancy. Many of the ideas appertaining to the origin of the solar system were as yet under the influence of some untenable theories.

Oparin assumed in his paper that the earth first appeared in a high temperature gas and dust form, and then cooled down and solidified. Part of his early work suffered from changes in our knowledge of the stars and planets, and particularly of the inner structure of the earth. The switch from a concept of an earth primarily at a high temperature to one assembled from cosmic dust with a surface at least always cool has been the biggest change. Nevertheless, the difference in these theories on the origin of life proved to be relatively small, and this keen young Russian scientist thus produced a paper that contained ideas for a new programme in chemical and biological research which inspired the work of many other scientists. This work began in 1952, the first results being published a year later.

The early stage in organic chemical evolution was the preparation for, rather than the origin of, life. Starting where there was sufficient dust and gas to act as raw material, it continued along normal chemical lines, not in themselves involving any new principles of chemistry and physics. A process of selection must have already operated among these first newly formed molecules pre-selecting those that would at a later stage become a part of living things.

The Elements of Life and its Environmental Needs

LIFE AS WE KNOW IT on earth consists essentially of water containing a variety of highly complex carbon and nitrogen compounds. The four principal elements which make up the vast bulk of organic matter on our planet are hydrogen, carbon, oxygen and nitrogen. Phosphorus, magnesium, sulphur, potassium, calcium and traces of other elements are also present. The carbon atom is unique because of the many and complex compounds it can form with other carbon atoms, and with atoms of hydrogen, oxygen and nitrogen. Organic chemistry has thus been described both as the chemistry of substances derived from living creatures and plants, and as the chemistry of the compounds of carbon. Over a million organic compounds are known and the number identified is constantly increasing.

Water is the principal matrix of life on earth and it can reasonably be assumed that this has always been the case. Thus the physical states of water, and therefore its temperature, also fix the conditions under which life could start and develop. They lie approximately between 10 and 40 degrees centigrade.

This in turn severely limits the size of a planetary body capable of sustaining life. Such a planet would have to be large enough to ensure that its gravitational attraction will be sufficient to prevent the escape of the water molecules into outer space. In addition, any such planet would have to be between 100 and 200 million miles distant from our sun. On a closer planet the water might escape into space. On a further one the temperature would remain below 0°C and any water would freeze. In our own planetary system only the earth satisfies the conditions which allow the planet to retain liquid water. On Mars night temperatures drop below freezing and so far no evidence of water has been found there. Recent close range photographs and television transmissions show its surface to be desolate, rather like the moon's with large numbers of circular craters.

The earth is the only planet in our solar system with an atmosphere in which oxygen plays a dominating role. Both Oparin and Haldane realized that the present concentration of oxygen in our atmosphere must prove fatal to any form of spontaneous generation. The present concentration of free oxygen in the earth's atmosphere simply means that if life arose spontaneously anywhere in the world today, in the form of some biochemical compound, it would be quickly eaten by other more highly developed forms of life and be ultimately oxidized before it had a chance to develop further into higher organisms.

Much evidence exists that molecular oxygen in our atmosphere is itself the product of life. It is formed by plants by the light-requiring process of photosynthesis. This is the process occurring in all green plants by which light energy is used to fix carbon dioxide from the atmosphere in the form of carbohydrates. It is achieved by a complex series of chemical reactions involving many intermediate stages, but the outcome of these chemical reactions can be represented in a simplified form, thus:

$$\text{Water} + \text{Carbon dioxide} + \text{Light Energy} \xrightarrow[\text{Enzymes}]{\text{Chlorophyll}} \begin{array}{l}\text{Carbohydrates} \\ + \text{Oxygen.}\end{array}$$

Not all animals require sunlight to survive, but none but a few anaerobic organisms can exist without oxygen. We can see from the equation above that plants do not require oxygen but in fact they produce it during photosynthesis and they cannot grow for long in the dark. Did animals or plants develop first? Or were the first organisms dependent on a form of metabolism different from either? This third possibility now seems the most likely.

A great deal of evidence derived from studies of the metabolism, cellular structure and life histories of both animals and plants leads to the conclusion that they are specialized descendants of an organism that could have resembled some of the bacteria of today, capable of carrying on simultaneously the functions of animals and plants.

By his studies of fermentations Pasteur showed a hundred years ago that life without oxygen was possible.

The term fermentation has been in use since about 1600, and denotes a decomposition of foodstuffs generally accompanied by the evolution of gas. The best known example is alcoholic fermentation, in which sugar is converted into alcohol and carbon dioxide. This conversion was established by Gay-Lussac in 1815 and can be described by the simplified equation:

$$\begin{array}{llll} C_6H_{12}O_6 & \longrightarrow 2CO_2 & + 2C_2H_5OH \\ \text{Sugar} & \longrightarrow \text{Carbon} & + \text{Alcohol} \\ & \quad\ \text{dioxide} & \end{array}$$

Pasteur linked fermentation and putrefaction as comparable processes, both representing decomposition of organic matter in the absence of air. Pasteur also recognized fermentation as a simpler though less efficient process than oxidation, which had probably evolved before it. The appearance of life on earth eventually prevented any successful repetition of a life-forming process for such newcomers would quickly be destroyed by already existing forms of life. But the absence of spontaneous generation today is no proof that it might not have occurred in a world where the atmosphere was of a different composition.

Nineteenth-century chemists and astronomers had no means of ascertaining what the earth's atmosphere might have been like thousands of millions of years ago. Today, spectroscopic analysis of the planets Jupiter and Saturn show that both have an oxygen-free atmosphere primarily composed of methane and ammonia. This provides supporting evidence that the earth, too, might at one time have had a similar atmosphere.

Life Needs Energy

As MORE EVIDENCE accumulates it becomes increasingly clear that the synthesis of such simple hydrides as methane and ammonia seems a normal occurrence in our part of the universe, and that their appearance is not confined to a particular place or a particular time. If the primitive earth, too, had such an atmosphere then ammonia, methane and water vapour could have reacted to form more complex compounds, thus paving the way to the synthesis of some basic biochemical substances. The energy necessary for these processes may have been largely supplied by the short-wave ultra-violet light from the sun. Ultra-violet rays are filtered out by the existing atmosphere owing to the presence of oxygen, which itself does not absorb these rays but is changed by them into the gas ozone in the upper regions of the atmosphere. The very fact that ultra-violet rays from the sun can today scarcely penetrate the earth's atmosphere is yet another probable reason why life does not now arise spontaneously on our planet.

The question inevitably arises: from where is derived the supply of energy to drive the various chemical reactions and transformations necessary for the origin of life?

Today, energy emanates from the constant turnover of a small fraction of the amount of solar radiation reaching the earth. This energy is turned by plants into food during photosynthesis. In former geological ages this same process produced vast extra energy stores in the form of coal and oil. But energy could not be stored in these forms until the highly complex molecules of chlorophyll had been evolved, which made the plant process of photosynthesis possible. Most scientists, therefore, believe that there must have been built up some original source of free energy in the form of energy-rich compounds before the usual energy-wasting process of life could begin.

This primary accumulation of energy-rich compounds must have been made possible in the first place by some form of radiation. The most important source of this radiation came direct from the sun, but a significant part could also have been played by the sun's hydrogen wind, or radioactive energy derived from radioactive elements that were, thousands of million years ago, far more active than they are today. The so-called 'primitive soup' thus formed in the waters of the earth had a kind of built-in free energy, and from it derived the first stage of the evolution of life. This was probably a process whereby the accumulated free energy was gradually converted into increasingly complicated molecular structures. Later, a new source of energy must have appeared, just sufficient to keep life in existence, allowing evolution to proceed. It is also possible that nature's first experiments with life were failures. Once, however, the process of photosynthesis, based on visible solar radiation, had been evolved in plants, ample energy became available, with the result that more energy was built into the organisms than could be used up in their own metabolism. Large amounts of this energy are released by the action of animals and the life processes of fungi, but enormous stores still exist in buried fossil fuels such as coal and oil.

Miller's Experiment

OPARIN held the idea that complex organic compounds such as the amino acids which form a basis for living organisms could have been synthesized by the combination of such simple substances as ammonia (NH_3), methane (CH_4), hydrogen (H_2) and water (H_2O) which are present in large quantities in the universe. This theory seemed eminently plausible, and in the 1950s a number of scientists set about to prove it by laboratory experiments.

Professor H. Urey, working at the University of Chicago, discussed this intriguing question with one of his students, Stanley Miller. In the spring of 1953 Miller published a paper reporting the results of his now classic experiment. His paper started as follows:

The idea that organic compounds that serve as the basis of life were formed when the earth had an atmosphere of methane, ammonia, water and hydrogen instead of carbon dioxide, nitrogen, oxygen and water was suggested by Oparin, and has been given emphasis recently by Urey and Bernal. In order to test this hypothesis, an apparatus was built to circulate CH_4 (methane), NH_3 (ammonia), H_2O (water) and H_2 (hydrogen) past an electric discharge. The resulting mixture has been tested for amino acids . . .

Miller conducted his experiments in an airtight apparatus of relatively simple design.

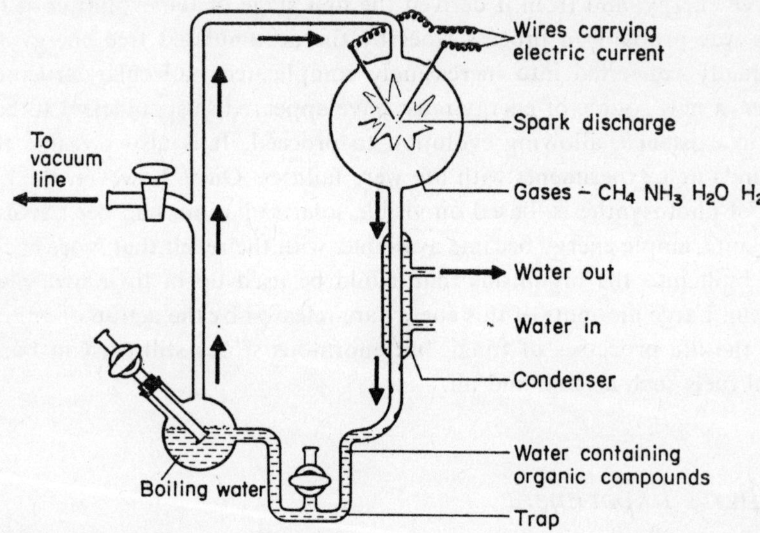

FIG. 17. A diagram of the spark-discharge apparatus used by S. Miller. (Adapted from Biological Sciences Curriculum Study, *Biological Science: Molecules to Man*, Houghton Mifflin Co.)

In it he placed the gases methane, ammonia and hydrogen, which he circulated past electric discharges from tungsten electrodes. Heat and water vapour were supplied by a container of boiling water connected to the apparatus. The steam from the container of boiling water mixed with the other gases in the large chamber where the electrical discharges between the electrodes took place. As the water vapour circulated it cooled and condensed, carrying with it some of the less volatile products of the experiment. Thus the conditions that may have been present in the early atmosphere were duplicated as accurately as possible.

Miller allowed the gases in the apparatus to circulate for a week and afterwards examined the accumulated liquid for its chemical content. A truly amazing variety of organic compounds had been formed from the simple starting materials, methane, ammonia, hydrogen and water. Among them were found a number of amino acids, the building blocks of proteins. Now there are only about twenty amino acids found in living matter. When

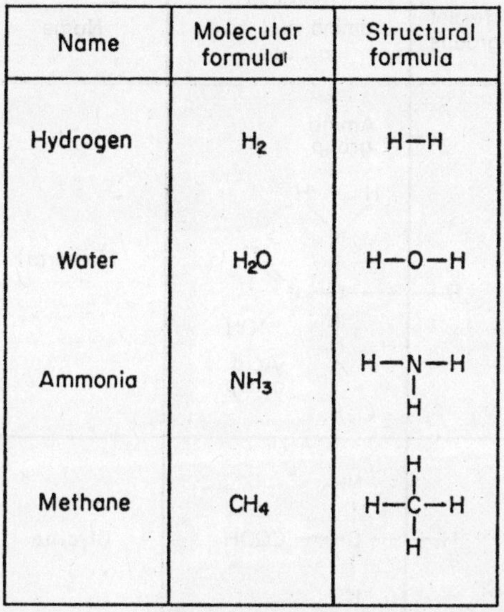

Name	Molecular formula	Structural formula
Hydrogen	H_2	H—H
Water	H_2O	H—O—H
Ammonia	NH_3	H—N(—H)—H
Methane	CH_4	H—C(—H)(—H)—H

FIG. 18. Molecular and structural formulae of hydrogen and some of its compounds used by S. Miller in his spark-discharge apparatus.

joined together in chains by so-called peptide bonds, they can form giant molecules consisting of as many as three thousand amino acid units in one molecule of protein. Proteins have been found in every form of life that has been analysed. They constitute a vital part of all living matter, and it is certain that they were an intrinsic part of early forms of life.

In the last twenty years a great deal of work has been done to elucidate the structure of proteins so as to obtain a clearer picture of how the large number of amino acids might be arranged in these giant molecules. From this work one thing has clearly emerged—the sequence of those twenty-odd amino acids in any one protein is enormously complicated. At first, the picture was one of random confusion, but then scientists discovered that much could be learned if the chemical structure of a number of proteins could be compared. Indeed, such comparison may show us the way back to simpler substances which preceded the complex protein molecules existing now. Scientists working in this field are beginning to see what kind of biochemical substances and reactions preceded those of present day biochemistry.

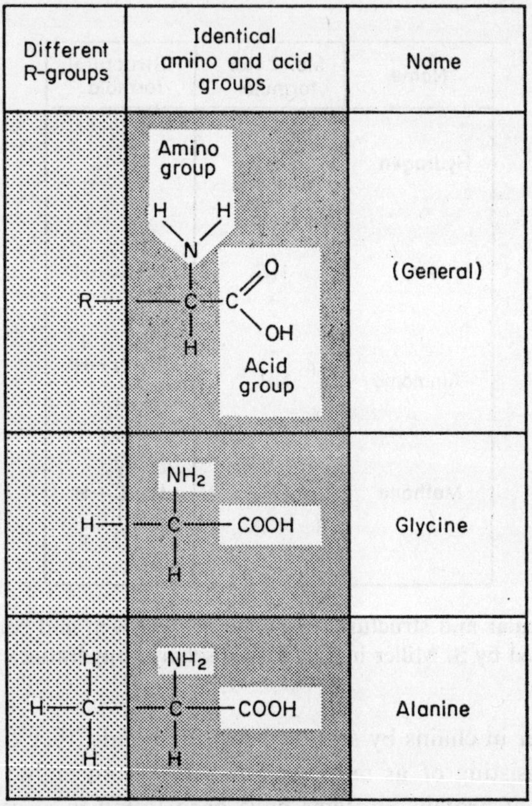

Different R-groups	Identical amino and acid groups	Name
R—	Amino group: H, H, N; —C—C(=O)OH, H; Acid group	(General)
H—	NH_2, —C—COOH, H	Glycine
H—C—(H)(H)	NH_2, —C—COOH, H	Alanine

FIG. 19. The structural formulae of a general and two specific amino acids. (Adapted from Biological Sciences Curriculum Study, *Biological Science: Molecules to Man*, Houghton Mifflin Co.)

Miller's experiment demonstrated that amino acids, the building blocks of proteins and thus an intrinsic part of all modern forms of life, could have been formed under the conditions present upon the primitive earth.

Some have argued that the very nature of powerful, ultra-violet radiation so destructive to life would have destroyed newly made organic molecules as fast as they were formed, and this objection is, indeed, based on sound reasoning.

One way in which these early organic products may have been protected from the harmful effects of ultra-violet radiation is by entrapment in the spaces between clay and sand particles on shores and river estuaries.

From Amino Acids to Proteins

CHEMISTS had known for some time that proteins are formed by biochemical reactions in living organisms. In fact any form of life acts like a chemical factory where those substances which are taken in as food are broken down into simpler constituents to be rebuilt by highly complex chemical reactions into the various constituents of the living organism. The question was now posed: How were the first protein molecules formed before life began?

Dr Sydney W. Fox, Director of the Institute of Molecular Evolution at the University of Miami, set about investigating this problem in the laboratory. Fox based his work on the results of Miller's experiments. In a paper presented by him in 1957 he had this to say:

> Working with an atmosphere consisting of ammonia, methane and water, Miller was notably able to produce amino acids by electrical discharge . . .
>
> The biochemical distance from such organic compounds as amino acids to the origin of life must be quite large.
>
> The work to be described in this paper, however, began with an attempt to understand only the pre-biochemical origin of protein. The experiment yielded a succession of unexpected results and stimuli for new experiments. The embarkation point for the experiment to be described was the hypothesis that peptide bonds might be formed at temperatures elevated enough to evaporate water.

Fox's reference to the formation of peptide bonds meant that he believed that by heating a mixture of amino acids to a temperature high enough to drive off water, bonds might be formed between the amino acid molecules, thus joining them together to form the much larger protein molecules.

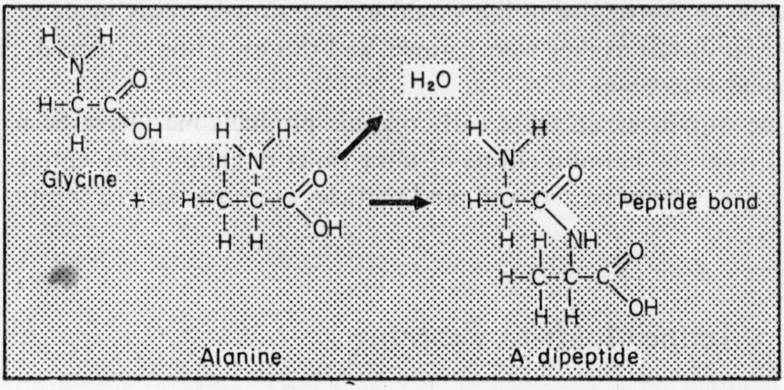

FIG. 20. The formation of a peptide bond. (Adapted from Biological Sciences Curriculum Study, *Biological Science: Molecules to Man*, Houghton Mifflin Co.)

Accordingly, Fox heated a dry mixture of amino acids to temperatures of 160–210°C for periods of half an hour to three hours. After the substance had cooled analysis revealed that many of the amino acids had joined together to form larger and more complex molecules. The water which had been given off during the experiment had evaporated. These more complex molecules he called protenoids because they were very similar to proteins. His experiments did throw some light on the puzzling question as to how proteins might have originated before there were living organisms capable of synthesizing them.

Professor J. D. Bernal of London University has suggested yet another way by which small organic molecules might have joined together (polymerized) to form more complex ones.

He put forward the theory that the growth of molecular particles took place on tiny particles of clay and mineral crystals which could be found in a finely divided state on beaches and particularly in river estuaries of the primitive earth. His theory is that the origin of life did not occur in open water but essentially in mud and soil. Most of the soil we know today is organically produced by the decay of animal and vegetable matter, but the primitive soil in a world where life did not as yet exist must have been derived purely from the various products of rock. Experiments have shown that simple organic compounds (monomers) are specifically attached to the surface of clay minerals where they occur sandwiched between layers of silicate. Many oil deposits being worked today are thought to be initially of this clay-absorbed nature.

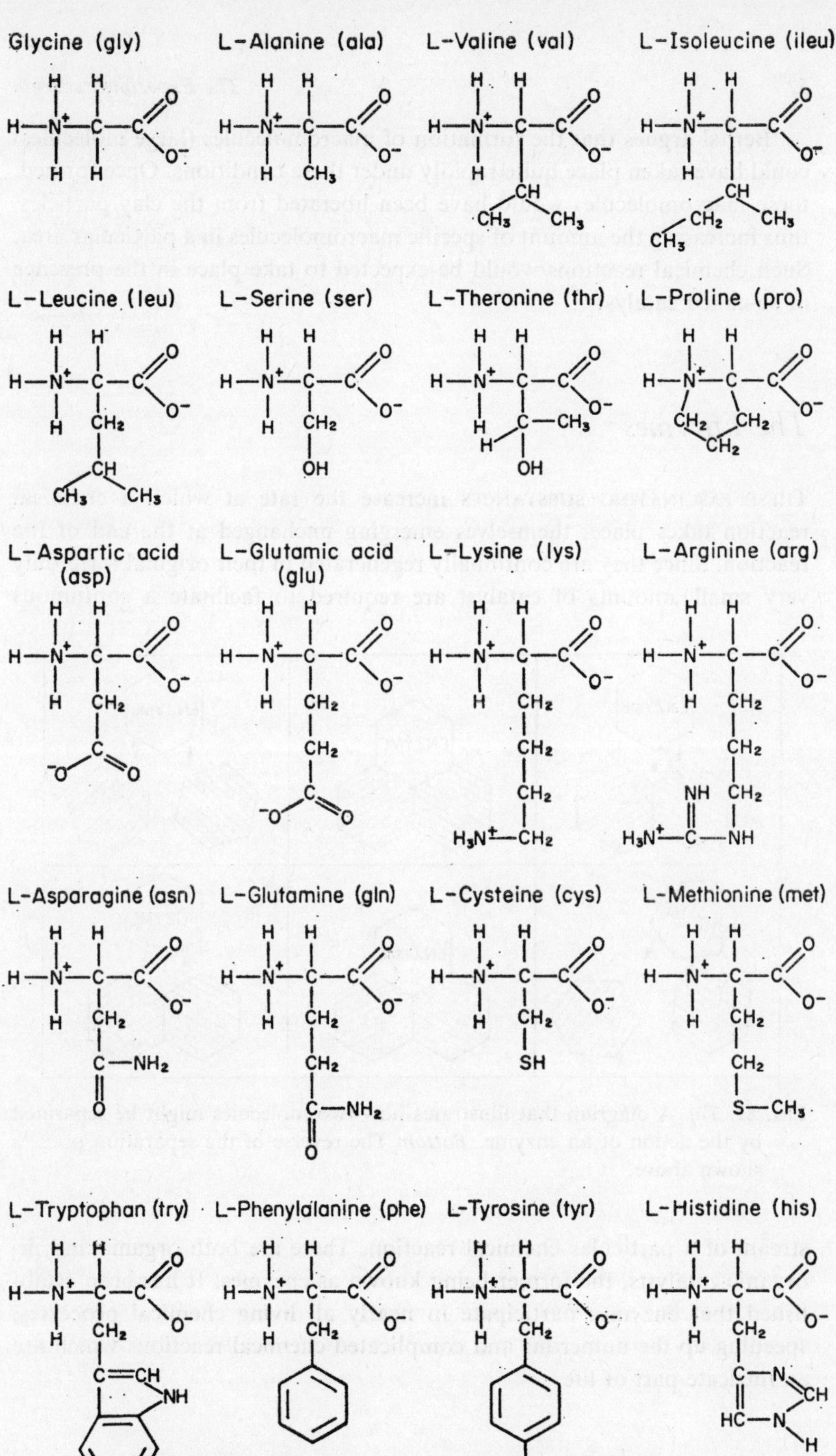

FIG. 21. The twenty common amino acids found in proteins.

Bernal argues that the formation of macromolecules (large molecules) could have taken place quite rapidly under these conditions. Once formed, these macromolecules would have been liberated from the clay particles, thus increasing the amount of specific macromolecules in a particular area. Such chemical reactions would be expected to take place in the presence of so-called catalysts.

The Enzymes

THESE FASCINATING SUBSTANCES increase the rate at which a chemical reaction takes place, themselves emerging unchanged at the end of the reaction. Since they are continually regenerated in their original form only very small amounts of catalyst are required to facilitate a continuous

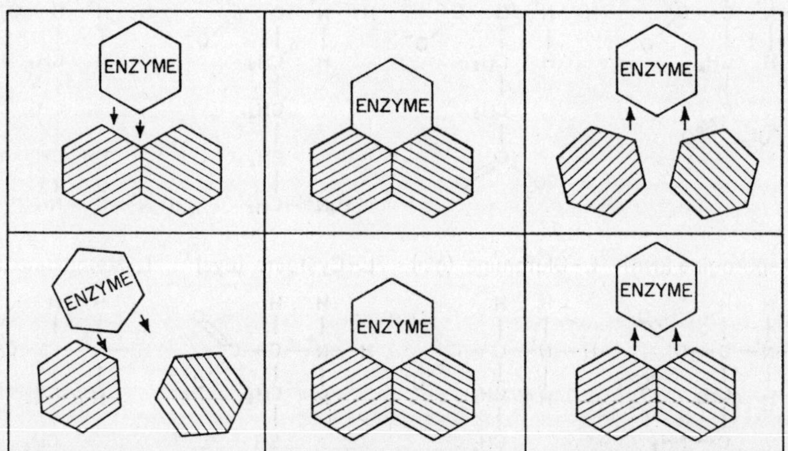

FIG. 22. *Top* A diagram that illustrates how two molecules might be separated by the action of an enzyme. *Bottom* The reverse of the separation process shown above.

stream of a particular chemical reaction. There are both organic and inorganic catalysts, the former being known as enzymes. It has been established that enzymes participate in nearly all living chemical processes, speeding up the numerous and complicated chemical reactions which are an intricate part of life.

Enzymes play such a dominant role in all life processes that it is difficult to imagine how life ever managed without them. Chemically they are proteins of a highly complex nature. The formation of a particular kind of enzyme depends on the presence of other enzymes, and we are therefore again faced with the problem of how enzymes could have been formed in the first place. If one remembers, however, that the enzyme is only a catalyst which can do no more than change the rate of a chemical reaction that would have happened in any case, only at a much slower rate, the problem no longer looks insurmountable. It simply means that at first many of the chemical reactions involving life processes proceeded very slowly until, by trial and error, those catalysts came into being that today speed life along its diverse paths.

The enzyme catalase occurs in plant and animal tissue, and decomposes hydrogen peroxide into oxygen and water.

Catalase is present in the blood of human beings and anyone who has cut a finger and applied hydrogen peroxide to the wound can see this amply illustrated by the frothy foam formed by tiny oxygen bubbles thus liberated. Catalase acts purely as a catalyst in this case, and is not itself used up in the process.

Alike Substances Tend to Aggregate

EVERYWHERE in the earth's crust physical and chemical processes are taking place which tend to bring together like elements and compounds. The relative abundance of the various elements in the earth's crust varies enormously. Thus, it contains 27·7 per cent silicon, 5 per cent iron, 0·000006 per cent silver, and only 0·0000005 per cent gold. Even the rare elements have tended to concentrate in certain minerals. This has enabled man to make use of some of them since earliest times. An example from industry serves well to illustrate this tendency to concentrate. For the generation of steam power-stations use water of great purity in their boilers, containing only a few parts per million of any kind of impurity. Yet the tiny quantities of copper derived from the wearing away of the washers of the pumps which circulate the water are able to accumulate to form lumps of pure copper weighing several pounds. It is reasonable to suppose that similar selections and segregations took place among organic molecules in the 'primitive soup'.

The Colloids and Coacervates

THE OCCURRENCE of colloidal substances was an idea Oparin introduced into his studies at an early stage. It was based on the old suggestion that primitive life occurred in some sort of jelly at the bottom of the sea.

Colloids are substances consisting of very tiny particles which remain separated to a greater or lesser extent by a solvent. Many colloids occur naturally; others can be prepared by breaking large pieces of material into smaller ones. There are three important types of colloids: (a) small solid particles which have the same internal structure as the solid from which they are derived (a suspension of sulphur in water is an example); (b) an aggregate of smaller molecules which have adhered (a familiar example is a soap solution); (c) the type which interests us most—large organic molecules that are of colloidal size (1/1,000,000—1/1000 millimetre), such as the proteins.

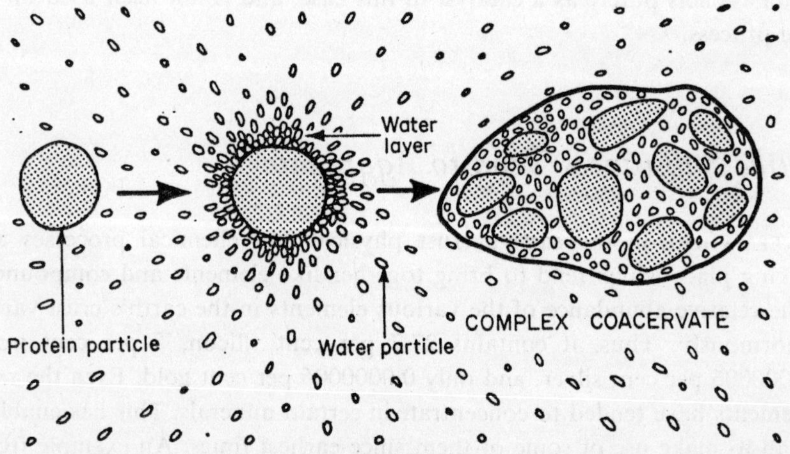

FIG. 23. The formation of a coacervate droplet. (Adapted from Biological Sciences Curriculum Study, *Biological Science: Molecules to Man*, Houghton Mifflin Co.)

If such organic compounds as amino acids and proteins were formed in the pre-life era on the earth, colloidal solutions of proteins most likely became distributed in the oceans, lakes and water holes of our planet. How then might these compounds come together to form still more complex structures? Oparin attempted to answer this question by drawing attention to a peculiar group of droplets named coacervates, derived from the Latin word *acervatus* meaning heaped or clustered. Coacervates, then, are clusters of proteins or protein-like substances held together in small droplets within a liquid medium. The formation of coacervates is dependent upon the fact that when proteins are dissolved in water, parts of the protein molecules become ionized. This means that they acquire an electric charge. Such charged protein molecules attract an array of water molecules around themselves which form an organized sphere. A coacervate results when, within a sphere of water molecules, a cluster of protein ions is trapped.

In Oparin's own words:

> The phenomenon of coacervation is particularly interesting from our point of view, in that, during the process of evolution of organic substances it must have been a powerful means of concentrating compounds of high molecular weight in particular protein-like substances dissolved in the hydrosphere (free water on the earth surface).

The idea of a chemical evolution, and even of chemical struggle for existence, is apparent in Oparin's writings, and was followed up considerably in his later work.

For example, one particular aggregate of molecules might well have been more successful in assimilating other chemicals from its environment than a similar aggregate. Thus at the expense of the latter the former would grow faster and more successfully. This process would eventually lead to the elimination of the less successful ones. It should be said that, to date, practically nothing is known about the epoch of chemical evolution on our planet, and it is therefore largely a matter of speculation as to how this might have occurred.

Our experimentations and speculations on this subject probably bear only a remote relation to what actually happened three or four thousand million years ago, but this does not mean that a biochemical enquiry on a subject of such complexity is an idle waste of time. On the contrary, even if no immediate solution is at hand, other branches of science, like medicine and molecular biology, have indirectly benefited from such researches.

Reproduction and the Nucleic Acids

WE HAVE NOW enquired at some length into possible ways by which complex organic substances such as the proteins might have been formed in an as-yet-lifeless earth environment. Some have called this stage the evolution of metabolism. Yet during this period it is likely that complex reproductive mechanisms also evolved on which were based the continuation of all forms of life as we know them today. Indeed all life, be it animals, plants, bacteria or viruses, depends for its continuance on earth upon a process of reproduction which ensures that all the characters of a species are passed on from one generation to the next. In turn the reproduction of organisms depends upon the replication of certain molecules, the nucleic acids. Not all authorities agree that a coded nucleic acid—a macromolecule—was among the first biologically active substances on our planet. Yet their universal presence in all organisms and their fundamental role in all forms of reproduction indicate that they must have appeared very early on in the evolution of life.

Let us review how scientists became familiar with these remarkable substances.

It is true to say that all living organisms, with the exception of the viruses, are constructed from cells. Their existence had remained unknown until, in 1665, the Englishman Robert Hooke sharpened his penknife to cut and examine pieces of cork. His subsequent observations are best described in his own words:

> I took a good clear piece of cork and with a penknife sharpened as keen as a razor, I cut a piece of it off, and thereby left the surface of it exceedingly smooth, then examined it very diligently with a microscope, methought I could perceive it to appear a little porous; but I could not so plainly distinguish them as to be sure that they were pores . . . I with the same sharp penknife cut off from the former smooth surface an exceeding thin piece of it, and placing it on a black object plate . . . and casting the light on it with a deep plano-convex glass, I could exceedingly plainly perceive it to be all perforated and porous, much like a honeycomb, but that the pores of it were not regular . . . these pores, or cells, were not very deep, but consisted of a great many little boxes, separated out of one continued long pore by certain diaphragms . . .

Hooke thus named the fundamental unit of living matter the 'cell'. His investigations having been carried out on plant materials, Hooke had actually seen the cell walls which form the outer boundary of each cell. Normally plants have very distinct cell walls, easily visible under a microscope, whereas animal cells do not have such distinct cell walls. Both types, however, contain living material called cytoplasm. Nearly all cells contain a nucleus that exercises control over the functioning of the cell. The nucleus has become one of the most closely studied microstructures of organisms.

When Anton Leeuwenhoek, the Dutch naturalist, was examining a drop of salmon's blood under the microscope, he noticed that each blood corpuscle had what appeared to be a light centre. Although he did not realize it at the time, what he had actually seen was the nucleus of the blood cells. In mammals, however, red blood cells have no nucleus.

The suggestion that the cell is the basic unit of life, both in plants and animals, was made in 1839 by the two German scientists, Theodor Schwann (1810–82) and Matthias Jacob Schleiden (1804–81). A few years later the German botanist and pioneer microscopist, W. F. Hofmeister (1824–77), observed that before a cell divides into two its nucleus also divides to produce two daughter nuclei. He also noticed that before a cell division the nucleus changed its appearance.

Later the development of special dyes made it possible to stain the nucleus, which, in the dividing cell, transforms itself into sausage-shaped bodies now called the chromosomes. All animal and plant cells capable of cell division were found to contain chromosomes, and it was soon realized that the number of chromosomes is always identical in animals and plants of the same species. Thus a human cell possesses 46 chromosomes, a Cebus monkey 54, an onion 16, a fruit-fly 8, and *Radiolaria*, a marine protozoa, more than 800.

Scientists had suspected for some time that the blueprints of heredity lurked in the chromosomes, but they had no idea what they looked like and from what they were made.

In 1903 Walter S. Sutton, working at Columbia University, suggested that the factors controlling heredity (genes) must be located in the reproductive cells, the egg and sperm, because these alone were able to pass hereditary material from parents to offspring.

But long before scientists began to associate the nucleus with the processes of heredity, Friedrich Miescher, a Swiss biochemist completing his studies in Strasbourg in 1869, attempted the chemical analysis of the nucleus. For a supply of nuclear substance he used the sperm of salmon, a rich source of nucleic acid, each sperm consisting almost entirely of nucleus. Miescher used a strong salt solution to dissolve the sperm heads, and then managed to precipitate the nuclear material by adding ordinary water. After much careful and exacting work, he succeeded in analysing the chemical elements present in the nucleus. He found carbon, oxygen, nitrogen and phosphorus, all the elements mentioned earlier as being important constituents of life. He called this new material 'nuclein' and, when later it was found to exhibit the chemical structure of an acid, it was named nucleic acid.

At first little notice was taken of Miescher's discovery, and for many years bottles of this gummy white powder grew dusty on laboratory shelves. Then chemists in Germany and America got down to the task of discovering how the atoms in this fascinating substance were arranged. Their experiments revealed that nucleic acid consisted of a very large number of atoms indeed. Compared with the molecules of such substances as water or table salt, nucleic acid was a veritable giant. They also found that it had a very long and thread-like structure. To this extremely complex substance they gave the equally complex name of deoxyribonucleic acid, thereafter known as DNA. Soon after the discovery of DNA another unique acid was found in the nucleus of yeast cells. This was named ribonucleic acid, shortened to RNA. For many years it was wrongly thought that RNA was confined solely to plants, but later scientists discovered that both DNA and RNA appeared to be constituents of most cells, animal *and* plant.

Although nucleic acid molecules are very large, they form an assemblage of only a few kinds of smaller molecules known as nucleotides. In both RNA and DNA each nucleotide consists of a phosphate, a sugar, and a nitrogen-containing molecule. DNA and RNA are named after the sugar molecules they contain. RNA contains the sugar ribose while DNA contains deoxyribose, a very similar sugar which differs from the ribose sugar only by having one oxygen atom missing.

At first nobody realized the widespread importance of these substances, but as more and more animal and plant cells were examined, it became apparent that they all contained some nucleic acid. Experiments began to reveal, to an as-yet-disbelieving scientific world, that DNA and RNA could be the very substances that exerted primary control over the basic processes of life in all living organisms. These discoveries, however, did little at first to solve the mystery. On the contrary, they caused more bewilderment than ever, for it seemed that DNA and RNA were somehow able to hold within their chemical structure all the instructions needed for the building and functioning of such diverse organisms as an elephant, a human, and the daisy in a lawn. To make a working drawing of a human being down to the finest detail would take dozens of sheets of paper. How then was it possible to concentrate in a tiny piece of matter, weighing no more than a four-hundred-millionth of a milligram, this vast amount of information, and how could DNA and RNA store and transmit the thousands of different master-plans necessary for the construction of all living things?

In the early 1950s feverish activity began in biochemical laboratories all over the world to find an answer to these remarkable questions. The optimistic view was even expressed that the stage was now set for the final breakthrough to produce life in the laboratory.

In 1953, J. D. Watson and F. H. C. Crick, two scientists working at Cambridge, England, began to strip the secrets from the mysterious DNA. With the aid of a molecule building-set consisting of some wire and a set of coloured wooden and plastic balls, they began to construct a model of the DNA molecule according to the various facts so far discovered in

FIG. 24. The DNA molecule has the form of a helical ladder, according to the Watson-Crick model. (Adapted from Biological Sciences Curriculum Study, *Biological Science: Molecules to Man*, Houghton Mifflin Co.)

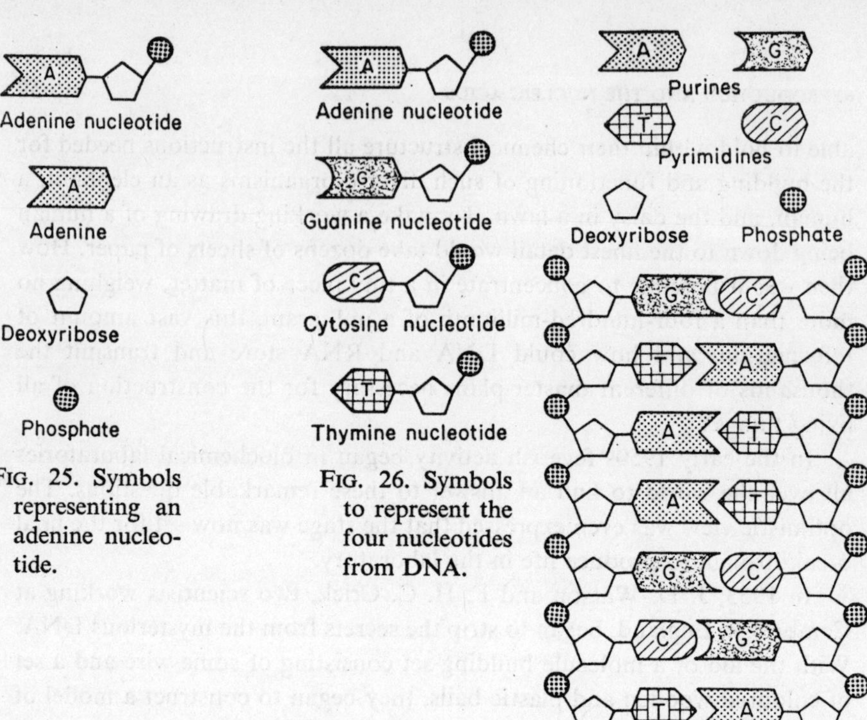

FIG. 25. Symbols representing an adenine nucleotide.

FIG. 26. Symbols to represent the four nucleotides from DNA.

FIG. 27. A symbolic representation of a small part of a DNA molecule, unspiralled to make it easier to study.

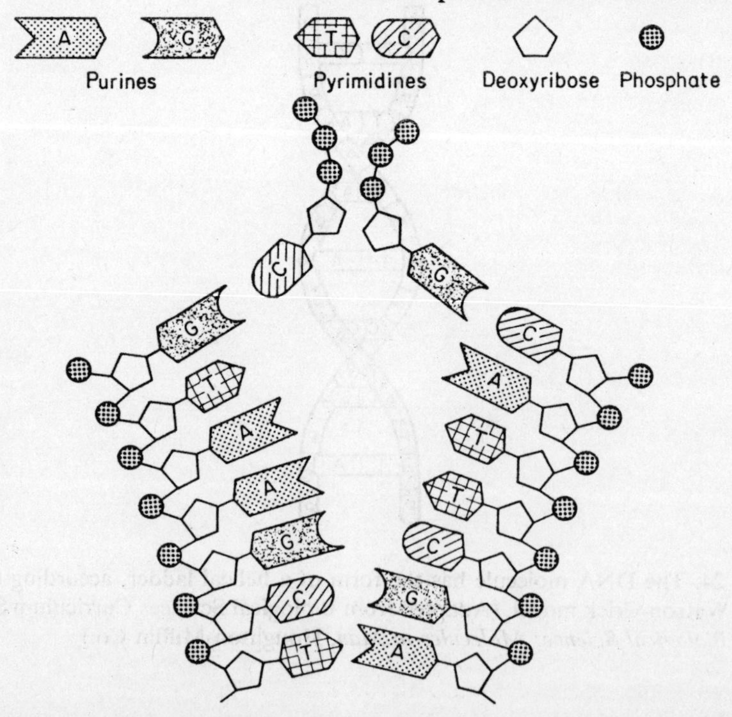

FIG. 28. The two strands of the DNA molecule begin to 'unzip' and new nucleotides of the proper kind, temporarily carrying extra energy-rich phosphate groups fall into place.

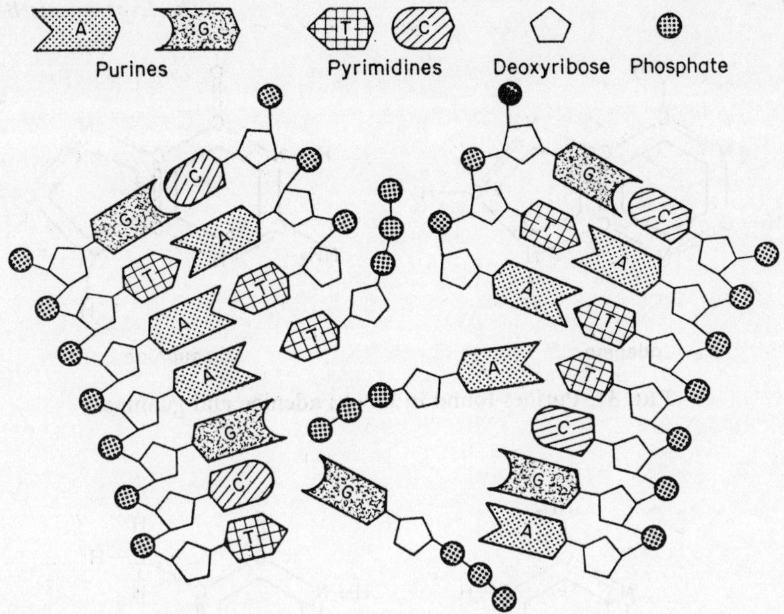

FIG. 29. The two original strands of DNA are completely separated and new nucleotides are being added.

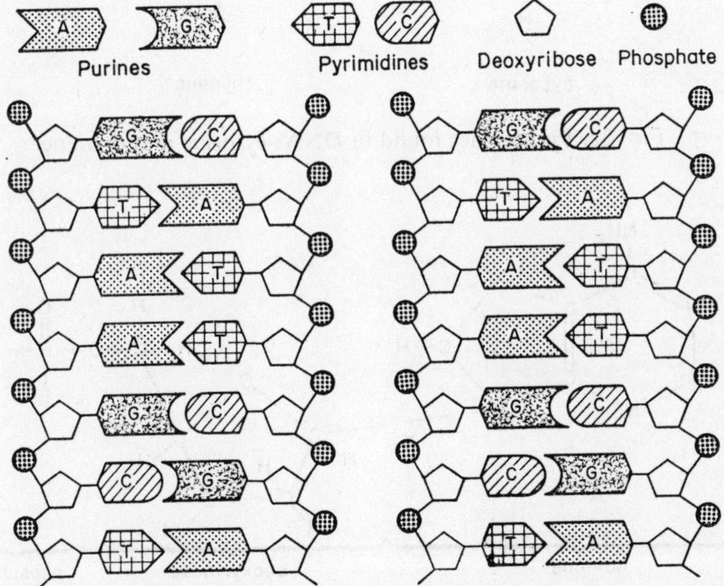

FIG. 30. The process of DNA duplication has now been completed. Each new nucleic acid molecule is exactly like the original one. (Adapted from Biological Sciences Curriculum Study, *Biological Science: Molecules to Man*, Houghton Mifflin Co.)

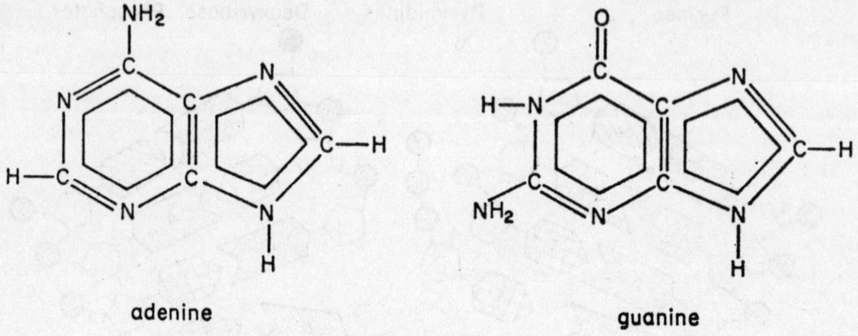

adenine guanine

FIG. 31. Purines found in DNA: adenine and guanine.

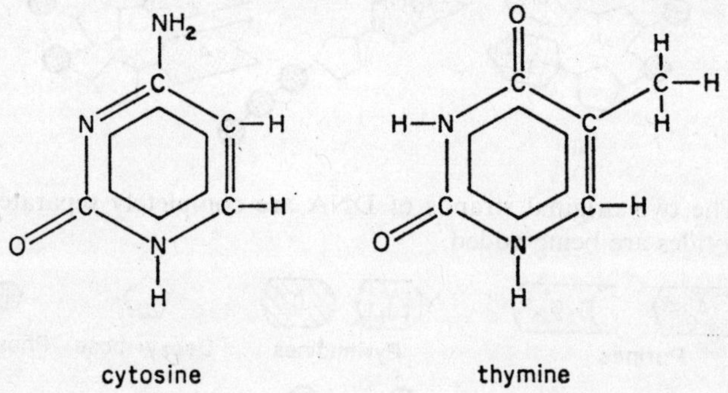

cytosine thymine

FIG. 32. Pyrimidines found in DNA: cytosine and thymine.

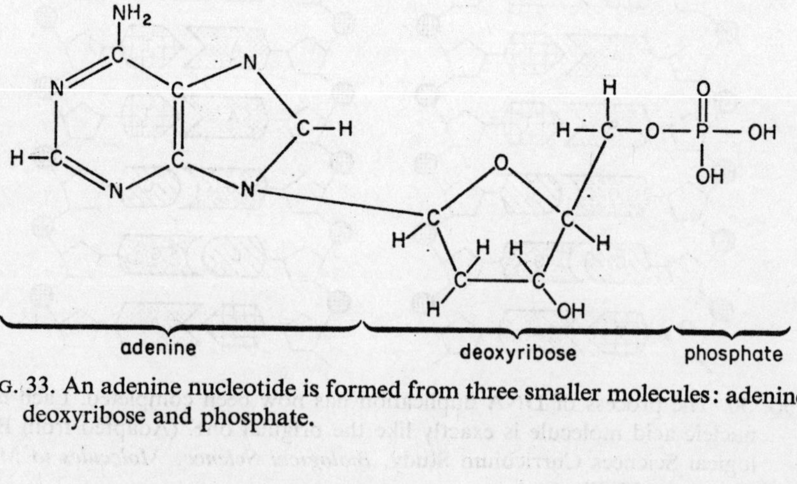

adenine deoxyribose phosphate

FIG. 33. An adenine nucleotide is formed from three smaller molecules: adenine, deoxyribose and phosphate.

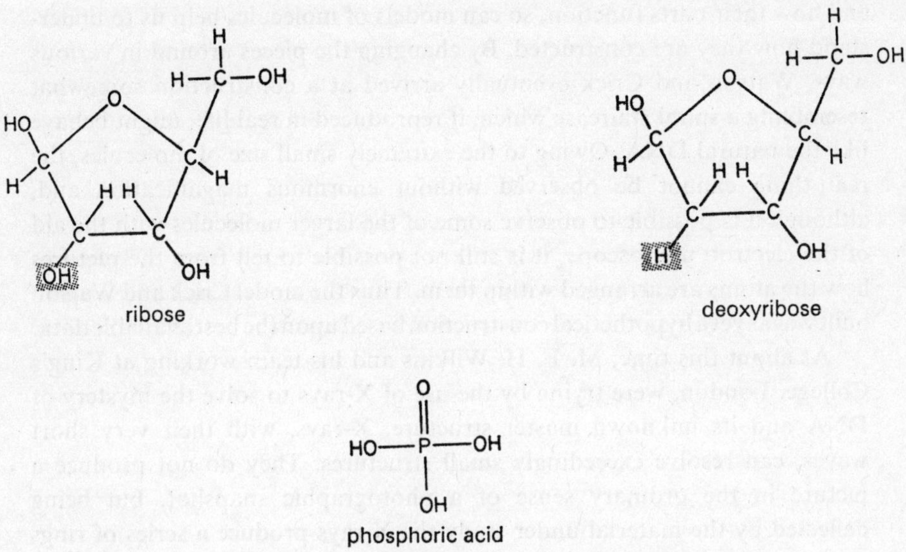

FIG. 34.

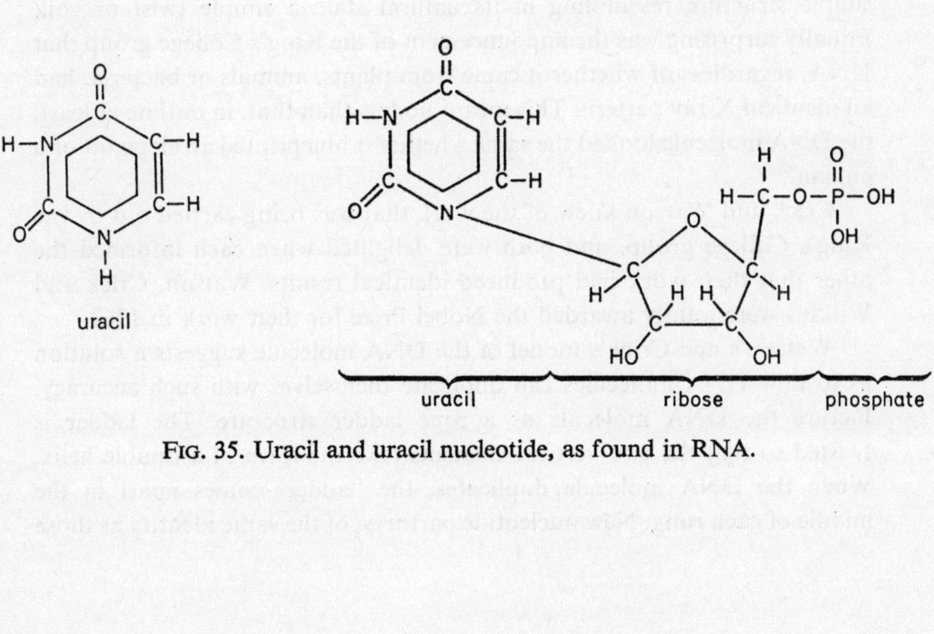

FIG. 35. Uracil and uracil nucleotide, as found in RNA.

connection with this substance. Just as the model of a ship, car or aeroplane helps us to understand how these mechanical contrivances are built and how their parts function, so can models of molecules help us to understand how they are constructed. By changing the pieces around in various ways, Watson and Crick eventually arrived at a construction somewhat resembling a spiral staircase which, if reproduced in real life, might behave like the natural DNA. Owing to the extremely small size of molecules, the real thing cannot be observed without enormous magnification and, although it is possible to observe some of the larger molecules with the aid of the electron microscope, it is still not possible to tell from the pictures how the atoms are arranged within them. Thus the model Crick and Watson built was as yet a hypothetical construction based upon the best available data.

At about this time, M. F. H. Wilkins and his team working at King's College, London, were trying by the use of X-rays to solve the mystery of DNA and its unknown master structure. X-rays, with their very short waves, can resolve exceedingly small structures. They do not produce a picture in the ordinary sense of a photographic snapshot, but being deflected by the material under study the X-rays produce a series of rings and patterns on a photographic plate which the expert is able to analyse by measuring angles and distances between the various dots and lines supplied.

Wilkins was as surprised by his results as were those who subsequently read his initial report, which simply stated that DNA was essentially a simple structure resembling in its natural state a simple twist or coil. Equally surprising was the announcement of the King's College group that DNA, regardless of whether it came from plants, animals or bacteria, had an identical X-ray pattern. This meant no less than that, in outline at least, the DNA molecule looked the same whether it blueprinted an elephant or a human.

Crick and Watson knew of the work that was being carried out by the King's College group, and both were delighted when each informed the other that their work had produced identical results. Watson, Crick and Wilkins were jointly awarded the Nobel Prize for their work in 1962.

Watson's and Crick's model of the DNA molecule suggests a solution as to how DNA molecules can duplicate themselves with such accuracy. Picture the DNA molecule as a rope ladder structure. The ladder is twisted so that the two vertical elements take the form of a double helix. When the DNA molecule duplicates, the 'ladder' comes apart in the middle of each rung. New nucleotide partners, of the same identity as those which have just been removed by parting, fall into place along each strand, once more forming a complete DNA molecule consisting of half the original molecule plus a corresponding set of new parts. In this way each of the two new molecules contains half the original molecule plus a set of new nucleotides. Thus both molecules remain identical.

The DNA molecules in a cell act as a reference library of instructions for the synthesis of the many different types of molecule required in a cell. Each form of life possesses its own distinctive set of proteins, and the DNA of an organism controls the manufacture of particular protein molecules. Instructions from the DNA molecules are transferred to messenger RNA molecules which then pass from the nucleus into the cytoplasm of the cell. There they become attached to small structures called ribosomes. They are located along the many folds of an inner network of membranes called the endoplasmic reticulum. It was found that the ribosomes contain most of the RNA of the cell and that newly formed proteins are associated with them. Proteins are thus synthesized on the patterns of the RNA molecules according to the original DNA instructions.

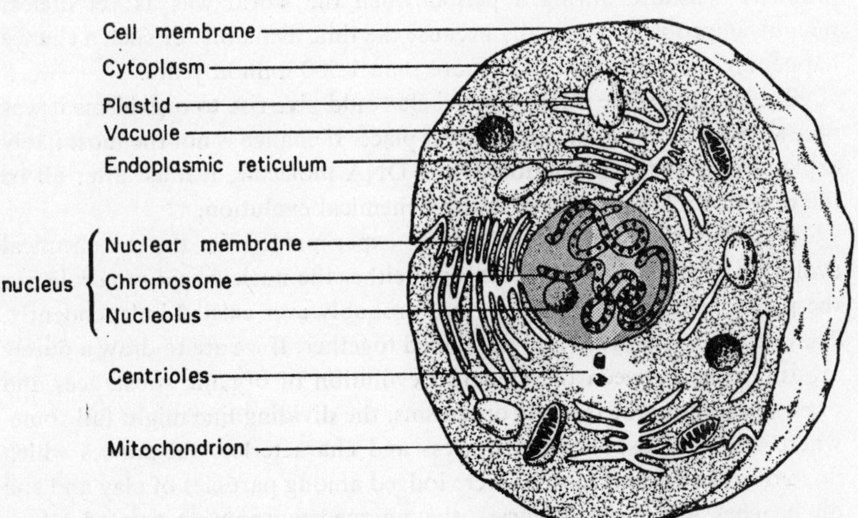

FIG. 36. The generalized cell as interpreted in the light of recent discoveries. (Adapted from Biological Sciences Curriculum Study, *Biological Science: Molecules to Man*, Houghton Mifflin Co.)

Experiments have shown that nucleic acid molecules can indefinitely replicate themselves by an automatic process. Thus the unravelling of the structure of nucleic acids provided at last a plausible model of a self-replicating molecule.

This explains also how it is possible that cells can divide indefinitely without losing their identity. During each cell division all the contained sets of instructions are passed on from one generation to the next. The nucleic acids DNA and RNA are either singly or together present in all forms of life, and among their functions is the vital control of heredity and the synthesis of proteins. This indicated to many workers that the nucleic acids must have been present at a very early stage in the evolution of life processes. Some even thought that the origin of life was synonymous with the self-reproducing molecule DNA.

The Nobel Laureates H. J. Muller and G. Wald as well as Dean Wooldridge, author of *The Machinery of Life*, invoke the laws of chance to explain the occurrence of very improbable events, for example the chance combination of the proper nucleotides in the proper number and arrangement to form a molecule of DNA. They observe that no matter how improbable an event (short of impossible) it is bound to occur given sufficient time. Yet the idea of a molecule of DNA arising by chance on a primitive seashore during a period when the world was as yet lifeless remains improbable especially because the time available for such a chance happening may not have been more than 1,500 million years.

But clearly before a DNA molecule could give rise to organisms it was necessary for it to be there in the first place. If chance is not the most likely explanation for the formation of the DNA molecule, it must after all be the result of a long chemical and biochemical evolution.

Thus by trial and error nature experimented in the biochemical evolution of the DNA molecule, but neither the nucleic acid molecule nor the protoplasm in which it operates probably ever existed independently. They must have evolved and developed together. If we are to draw a purely arbitrary line between the chemical evolution of organic substances and organic evolution of the first organisms, the dividing line might fall somewhere among those as yet shapeless and characterless substances which floated in the ancient seas, or were lodged among particles of clay and soil on beaches and river estuaries, the unique environment created where rivers entered the sea. Those which had acquired a mechanism for exact replication in the form of nucleic-acid-like molecules contained the rudiments of organisms as we know them today.

It is important to realize that any theory of the origin of life must incorporate in its stipulations mechanisms of molecular reproduction in order to enable any particular form of life to continue in existence. Even among the most primitive forms the concept of the survival of the fittest introduced by Charles Darwin must have already been operative.

The Synthesis of Biologically Active DNA

In 1967 a major breakthrough in the synthesis of biologically active DNA occurred at the Stanford University School of Medicine in California. Eleven years ago Dr Arthur Kornberg and his co-worker Dr Mehran Goulian set themselves the remote and difficult task of synthesizing in the laboratory living molecules of DNA. By painstaking work they eventually were able to synthesize in a test tube a DNA molecule which formed the genetic core of the bacterial virus ØX174. This meant, in fact, that Kornberg was able to synthesize a biologically active molecule from a complex series of organic substances which by themselves were not biologically active. Kornberg and his associates believe that these discoveries have opened the way to the copying of DNA molecules of other viruses and ultimately the much longer DNA of cells. In such experiments it should eventually become possible to change known forms of DNA by inserting alternative building blocks into their structure.

The work of Kornberg and his associates is not directly related to the question as to how life might first have started, yet it demonstrates that it is possible to synthesize biologically active molecules from biologically inactive organic substances.

How life originated is still very much a matter of speculation. A long period of development of carbon compounds leading to the formation of amino acids and proteins must have preceded the origin of life. The action of catalysts such as clay minerals probably played an important part. Nature's chemical experiments continued in a world as yet devoid of life. Molecules succeeded molecules in an endless stream of experimental successes and failures until eventually the first primitive forms of life were formed.

We have to traverse the immense span of time of what might have been more than 1,000 million years' development until at last we arrive at a period in the Pre-Cambrian era, some 2,800 million years ago, when nature's experiments with life left the first definite traces in the form of fossil records.

Some of the earliest such evidence has been found in the Bulawayan limestones of Rhodesia. It takes the form of nodules built up of thin layers of carbonate of lime estimated to be more than 2,800 million years old. These may have been formed by algae or bacteria for they are comparable with calcareous nodules formed at the present time by blue-green algae in the waters of streams and lakes.

Only a few undoubted traces of organisms have survived in Pre-Cambrian rocks, probably because most of the creatures living in those early seas were entirely soft bodied. Judging from fossil records life had become varied and highly organized long before the beginning of the Cambrian period 600 million years ago.

By the beginning of Cambrian times a considerable variety of invertebrates (animals without backbones) had come into existence in the sea, many of them with hard protective shells or skeletons; and unlike their Pre-Cambrian ancestors, they left unmistakable traces in the rocks in the form of fossils.

During later periods the succession of living forms became increasingly clearly imprinted as fossils in the rocks and strata of our planet.

3

The Experiments Continue

Evolution and Life

An Idea is Born

THE VERSATILE CAPABILITIES of matter lead us from the creation and motion of the stars in the universe to the intricate and complex chemical reactions from which living forms on our planet first originated. A seemingly endless succession of animals and plants winding their way through the geological eras of the earth have stimulated thinkers over the centuries to seek solutions for this happening. The question of accident or design was to become a fundamental source of discussion among scientists and laymen alike. The modern evolutionary theory in which it culminated explains in factual terms a process so remarkable that it took many years to make its intricacies as well as its basic simplicity fully understood. The essentially experimental nature of evolution will become clear as we investigate its diversities.

One of the obvious reasons why the process of evolution escaped notice for a long time was because so little change in nature was discernible during a single lifetime. Another difficulty was a gross misconception of the time available. Throughout recorded history stability seems to have been the rule, and clearly very little could have happened in the six thousand years allotted by interpretations of the Scriptures.

The true nature of fossils had already been appreciated by Leonardo da Vinci (1452–1519). He had suggested connections between them and the study of geological processes. Leonardo was far ahead of his time, however, when he applied himself to the study of sedimentary formations of fossiliferous rocks in many parts of his Italian homeland. Fossil deposits which he accepted as evidence of change in sea-level could not be accounted for by biblical descriptions of The Flood.

In 1785, James Hutton was among the first to hold the view that the earth had no beginning and was not likely to have an end. He observed that valleys might have originated from the action of streams, and that the waves of the ocean gradually changed and eroded the seashore. Thus, in very slow stages, the idea gained ground that changes on earth took place by a gradual process of evolution rather than by sudden acts of catastrophe.

The idea that life also had evolved took a much longer time to win support. The firmly rooted belief that the world came to exist in one grand act, and that all species had been created in an immutable form that could neither change into other species nor ever die out, was strongly built on the religious acceptance that each living form on earth had its place and purpose.

Eventually, in the eighteenth century, thinkers from many different disciplines began to question this idea. Fossil evidence demonstrated that the great mammoth (a fossil elephant) had lived in areas of the world where such animals had never been known to exist. Also, these creatures must have differed in appearance from the modern elephant, since the bone structure of their skeletons was similar but not identical to those of their modern counterpart. Fossil remains were also unearthed of a giant ground sloth, whose closest relative was the small tree sloth of South America. Soon questions began to be asked about the possible extinction of species, and the probability that one species may have evolved from another. Linnaeus (1708–78), the founder of the modern system of classification of plants and animals, still maintained that species were inviolate. The majority belief held at this time was that any change represented a degeneration from the ideal type to something inferior which in the end might lead to extinction. The notion that change might lead to improved or new species was not as yet generally held.

The French biologist Jean-Baptiste Lamarck (1744–1829) was the first biologist to propose a well-developed hypothesis explaining the means by which plants and animals evolve. His theory of evolution was published in 1809, the year Charles Darwin was born. Beginning with the idea that species could give rise to other species, he reasoned that a great change in the environment of an animal produced a need for change in its physical make-up. Lamarck assumed that the more a part of the body is used, the greater will be its consequent development and growth. Thus he would ascribe the long neck of a giraffe to the fact that this animal fed by stretching its neck to reach the foliage of trees. Conversely, he would argue that the disuse of an organ led to its reduction in size and eventual disappearance. He also postulated that any animal could pass on to its offspring those characters which, through use, had become enlarged or by disuse had grown smaller. This second assumption he called the 'law of inheritance of acquired characteristics'.

Lamarck's explanation of the process of evolution seemed convincing at the time, particularly because the first of his major assumptions is a valid one. Parts of animals do change as a result of use and disuse. Unfortunately, his second assumption was erroneous, as subsequent studies have repeatedly shown, for there is virtually no evidence to support the suggestion that acquired changes can be passed to offspring (or at least no method to indicate how this could be done).

Fifty years later, on 24 November 1859, Charles Darwin's monumental work, *On the Origin of Species by Means of Natural Selection*, was presented to the public. The whole edition of 1,250 copies of Darwin's book was sold out in a single day.

Born in Shrewsbury, Shropshire, Charles Robert Darwin (1809–82), the grandson of Erasmus Darwin, was educated at Shrewsbury School, and later went to Edinburgh University and studied medicine with little success. In 1828 he proceeded to Cambridge with the idea of becoming a clergyman. He actually took his degree in divinity in 1831, but immediately abandoned the idea of entering the Church and turned towards his real interest—the study of natural history. Helped and encouraged by his former teacher and friend, the botanist Henslow, twenty-two-year-old Darwin joined HMS *Beagle* in December 1831 to accompany the vessel on its surveying expedition to South America and the Pacific. Among his duties as ship's naturalist he was to study animals and plants and keep a written record of his observations.

During the course of his travels Darwin recorded a wealth of information. He soon became aware that three of his most important observations conflicted with the current theory of the immutability of species. The first was occasioned by his studies of the fauna of the Galapagos Islands. These volcanic islands lie on the Equator in the Pacific Ocean some five hundred miles west of South America, and three thousand miles east of Polynesia. Emerging from the sea by volcanic action more than a million years ago, they have never been connected with the mainland. The land animals that have made their home there must somehow have travelled over the sea, and only a very few species established themselves. In the course of time, these isolated islands have become populated by the descendants of the few creatures that did not perish on their hazardous journey.

The conditions are ideal for evolutionary studies—a natural laboratory set in the Pacific Ocean. The most striking animals on the islands are giant tortoises and great lizards. One kind of lizard is terrestrial, the other a marine kind that comes to land only to rest or mate. There are also flightless cormorants similar to those of the Chilean coast except for the reduction of wings, and many small birds, now known as Darwin's finches. These animals still live on the islands today but the direct inroads of man and of the egg-eating rats are rapidly destroying the best showpiece of evolution on earth.

Darwin's finches, small drab-looking birds, would in the ordinary way attract little attention except that there are as many as fourteen species, all similar to one another. We shall hear more about them later. Darwin was struck by the fact that if these birds had been specially created, there must surely have been the most extravagant expenditure of creative activity in the middle of the Pacific Ocean!

During his voyage on HMS *Beagle*, Darwin early visited the Cape Verde Islands near the Equator, off the coast of Africa. These islands are volcanic in origin, and the plants, insects, birds and other animals found there resemble closely those which exist on the nearby seashores of Africa. Several years later, when Darwin reached the Galapagos Islands, he was struck by the great similarity of the climate and terrain between them and the Cape Verde Islands. Surprisingly the plants and animals were totally different, more clearly resembling the species on the west coast of South America about six hundred miles to the east. The resemblance in each case to the flora and fauna of the nearby mainland made it reasonable to think that the organisms had originated there. Such an origin, followed by isolation and adaptation in their new homes, would best explain what had happened to these species. Many other kinds of geographic evidence pointed to dispersal and colonization followed by isolation and evolution in the new homes, but nothing was quite as dramatic as the evidence from the Oceanic Islands.

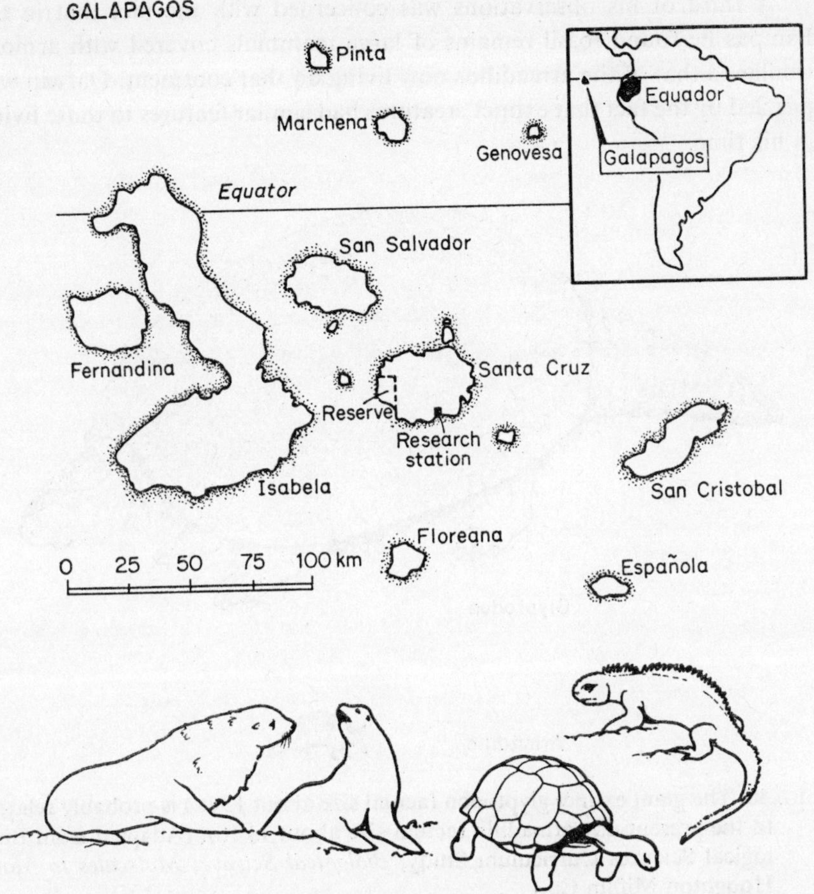

FIG. 37. The Galapagos Islands and some of their more important inhabitants, the sea lions, giant lizards, and giant tortoises (not drawn to scale).

During travels on the South American mainland Darwin made a second set of observations. Here he discovered that a number of species living in one particular area were replaced in neighbouring regions by other species that, though close in general resemblance, differed in some particular features. For example, the rheas (large ostrich-like birds) living near Buenos Aires were different from those he observed in the plains of Argentina at the southern tip of the continent. On the savannas of La Plata he was able to observe a rabbit-like animal built on the same plan as the South American type rodents, but which differed from those of North America and the Old World.

A third of his observations was concerned with the fact that in the Pampas he found fossil remains of large mammals covered with armour similar to that of the armadillos now living on that continent. Darwin was puzzled by the fact that extinct creatures had similar features to those living in his time.

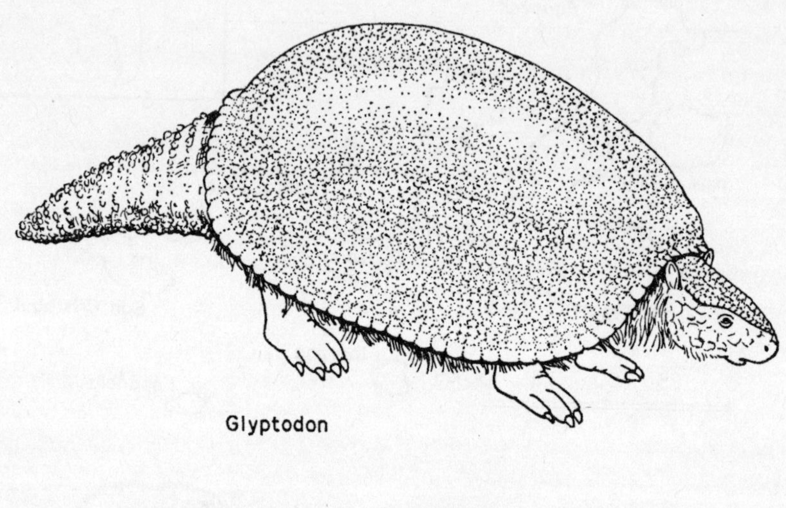

FIG. 38. The giant extinct glyptodon (actual size about 13 ft.) is probably related to the present day armadillo (actual size about 1½ ft.). (Adapted from Biological Sciences Curriculum Study, *Biological Science: Molecules to Man*, Houghton Mifflin Co.)

If one believed that species were immutable and had not changed since the day of their creation, there was no rational explanation for any of these observations. The matter would have to be referred to the realms of mystery and biblical creation. On the other hand, if species of animals or plants were subject to modification during descent through generations, and could even diverge into different lines of descent, simple and satisfactory answers to these questions could be found.

Darwin also studied variations in the breeds of domestic pigeons. Over many years pigeon breeders had developed numerous extreme types of birds all originating from the common pigeon so frequently seen in cities. Observations such as these, together with his interest in the origin of new varieties of cultivated plants, provided Darwin with further evidence of the role of natural selection in evolution.

In essence, evolution means that living things change: that species may alter and that some may die out and others come into existence. It means, therefore, that the plants and animals now inhabiting the earth were not the first plants and animals to exist, and that many of those that once flourished no longer exist.

Darwin's idea was that in each generation we should expect a slight increase among those who possess somewhat different characters, providing these gave some advantage. Here, then, is an experimental natural process that seems to develop by trial and error and involves the presence of individual hereditary variations, the tendency to increase in numbers and the struggle for survival. This whole process—especially the struggle for survival and the difference in rate of reproduction between the more favoured and the less favoured of the struggle—Darwin called natural selection. By this process species become better adapted to their environment in the particular habitat in which they live. Through the process of natural selection they acquire characteristics that improve their chances of survival.

This idea can be investigated by applying it to the English blackbird, a species found in nearly every garden. Each spring a pair of these birds rears two to three broods of up to four young. If all the offspring survived to reproduce, there would soon be too many blackbirds for a particular environment to support, and many of them would die.

In any brood some birds are stronger and better fitted to survive than others, and if there are many broods the differences and variations are correspondingly larger. Young blackbirds seem notoriously slow in spotting their enemies, and many of them therefore fall easy victims to attack. More alert birds, quicker to spot a stalking cat or hovering kestrel, are more likely to survive as juveniles than their duller brothers and sisters. Similarly, some young birds may be more favourably coloured in relation to their environment and thus escape the notice of their enemies.

Those that do survive will, of course, be the parents of the following generation. The new generation will tend to resemble the selected blackbirds who became their parents. The same will hold for the next generation, and the next and so on. The result will be a shift away from the norm of the original parent population until one day the shift away may be so great that the descendant birds can no longer be recognized as the same type as their forerunners. A new species of blackbirds will have emerged!

In 1858 Darwin received a manuscript from Alfred R. Wallace (1823–

1913), the English naturalist who was at the time on the island of Moluccas in the Malay archipelago. Darwin in England at once recognized his own theory in the essay. 'I never saw a more striking coincidence,' he wrote to Sir Charles Lyell the geologist. 'If Wallace had my manuscript sketch written out in 1842 he could not have made a better short abstract. Even his terms now stand as heads of my chapters.' Darwin was so impressed by Wallace's work that he almost yielded to him the honour of being the first man to announce the theory of the origin of species by natural selection. However, under the advice of Lyell and Sir Joseph Hooker, the two papers were presented under joint authorship at a meeting of the Linnean Society on 1 July 1858, using the title: 'On the Tendency of Species to form Varieties, and On the Perpetuation of Varieties and Species by Natural Means of Selection'. As previously mentioned, Darwin published his great work, *On the Origin of Species by Means of Natural Selection*, in November 1859. This massive treatise contained abundant evidence that animal and plant species had undergone a long process of change.

Darwin's work was received with enthusiasm by most of his friends and colleagues, who admired its logical approach. But when Darwin extended his ideas on evolution to the descent of man he encountered bitter and violent controversies with the established Church and a bitter war developed between Science and Religion concerning evolution. Feelings on this subject ran so high that soon afterwards Sir Richard Owen and Bishop Wilberforce set out to reduce Darwin's theory to ridicule at the British Association meeting at Oxford in June 1860. At this meeting Sir Richard Owen made the rash assertion that a gorilla's brain shows greater differences from the brain of a man than it does from the brain of the lowest apes. This statement was flatly contradicted by T. H. Huxley, the zoologist. Later during the meeting, when Bishop Wilberforce began his speech, feelings were running high. Churchmen everywhere felt that the doctrines of the Christian religion, based on the literal interpretations of the Bible, were being threatened by the arrogance of science. Accordingly, Wilberforce openly told his friends that he now intended 'to smash Darwin'.

He made an eloquent speech, and carried away by his enthusiasm he turned to Huxley and scornfully asked him: 'Is it on your grandfather's or your grandmother's side that the ape ancestry comes in?' His speech ended by the proclamation that Darwin's theory was contrary to the revelations of God in the Scriptures. He was wildly cheered by his audience, and the meeting would have ended in his favour if many of the students had not clamoured for Huxley's reply. After some persuasion Huxley rose and made his celebrated retort:

'I asserted, and I repeat, that a man has no reason to be ashamed of having an ape for his grandfather. If there were an ancestor whom I should feel shame in recalling it would be a man of restless and versatile intellect who, not content with success in his own sphere of activity, plunges into scientific questions with which he has no real acquaintance, only to obscure them by aimless rhetoric and distract the attention of his hearers from the point of issue by digressions and appeals to religious prejudice.'

The meeting ended with the final consensus of opinion being overwhelmingly on the side of science, and the great new concept of evolution. But the battle was not over.

During later years, although his correspondence shows that he was essentially a religious man, Darwin was denounced from the pulpits of almost every church in his homeland to the end of his days. He was never given the knighthood he deserved.

At one time, he wrote: 'I have never been an Atheist in the sense of denying the existence of a god'; and, 'I think Agnostic (a word coined by Huxley) would be the more correct description of my state of mind.'

Darwin's main concern was to search for the truth in nature. He felt that both scientific and religious opinion must accept and respect the evidence of established fact.

The Synthetic Theory

A THEORY as important and fundamental as the evolutionary theory itself stimulated scientists in many disciplines to a tremendous research effort. There were a host of questions to be answered.

Were evolutionary changes slow or rapid? Did they come about gradually or in jumps? Did evolution proceed by chance or was it directional?

Evidence for directional evolution has been eagerly sought by those seeking explanations in terms of man's supreme position in the universe. How was change accomplished, and what mechanisms were involved in the organism? Is the concept of natural selection valid, or might there be some guiding force, or vitalistic drive? Both Lamarck and Darwin recognized the process of evolution but each tried to explain its operation in different ways.

The basic idea of evolution, which is the primary theory, must be distinguished from the many related theories which are secondary. This division has a most important aspect, namely that the primary theory is assumed to be valid in any of the secondary theories.

Major theories are always complex and draw supporting evidence from widely differing fields. Today, the theory of evolution draws its chief supporting evidence from the primary sources of geology and biology, the latter including the categories of genetics, ecology, ethology, physiology and comparative anatomy. The bridge between biology and geology is palaeontology which supplies the time-scale and history. Chemistry and physics are, of course, basic sciences closely linked to modern biological studies through the field of molecular biology.

Each of these basic sciences has its own particular theories, many of which lend support to the primary theory of evolution. As many of the pieces of the puzzle fell into place a more elaborate evolutionary theory was developed. One of its principal spokesmen was the American palaeontologist, George Gaylord Simpson. It has been called the synthetic theory because it draws its evidence from a wide field of different disciplines.

Today the synthetic theory is the most widely accepted of all evolutionary theories. It is based in its fundamentals on Darwin's theory of natural selection, but the total structure is far more elaborate, being composed of many accessory theories intricately interwoven and interdependent.

Darwin considered natural selection the important driving force of evolution, but he, too, erroneously accepted the inheritance of acquired modifications as an important subsidiary element, a view which was shared by most of Darwin's immediate followers. Thus, for example, the dark skin of the Negro was considered to be due to the tanning of the skin by exposure to intense sunlight over many generations. Similarly, the absence of eyes in many subterranean animals was ascribed to inheritance of the effects of long-continued disuse of the eyes owing to their life in darkness. In 1892 the German biologist A. Weismann (1837–1914) challenged, by a somewhat crude experiment, the assumption that acquired characters are inherited. He cut off the tails of new-born mice in a series of successive generations, only to find that the twenty-first generation had tails just as long as the first generation of mice.

Darwin also proposed the theory which assumed that all organs of the body produced diminutive vestiges of themselves which he called gemmules. These he thought were shed into the blood stream and transported by blood to the sex glands where the gemmules of different organs are assembled to form sex cells. This ingenious idea was subsequently disproved by further knowledge of the mechanism of inheritance.

The Theory of Heredity

DARWIN'S lack of knowledge of heredity weakened the explanations he advanced during his lifetime. He was convinced that differences could be inherited, but he was unfamiliar with the way this process worked. He died ignorant of the fact that during his lifetime the Austrian monk Grégor Johann Mendel (1822–84) had by painstaking experiments pioneered the way to the discovery of the basic principles of heredity and subsequently the science of genetics.

Mendel began his work in the small garden of his monastery in 1856. He succeeded where others had failed, and this success was in part due to his training as a mathematician as well as a biologist. He crossed varieties of garden peas that had, under his observation, maintained constant differences in such single alternate characters as tallness and dwarflike growth and presence or absence of colour in blossoms, leaf axils and seed coats. He studied large numbers of offspring, and was able to use his mathematical knowledge to introduce the concept of probability. This was particularly important because in Mendel's time nothing was known about chromosomes or the process of cell-division, nor was the process of fertilization, the union of egg-cell and sperm (gametes), fully understood. Indeed, many years were to pass before these biological phenomena were recognized. The principles Mendel was able to establish were based entirely on his observations from his breeding experiments and these were in no way related to an understanding of the changes that were taking place in the cells.

One of his many experiments was made with tall and dwarf pea plants. When crossing these two varieties he found that in the first generation (F_1) only tall plants resulted. Tallness seemed to dominate over the alternate trait of dwarfness. Mendel, therefore, called a trait dominant if it appeared with no exception in the offspring of parents with contrasting traits.

Mendel's next experiment was to allow the first generation plants to pollinate themselves. Here a surprise awaited him, for both tall and dwarf plants reappeared in the same constant proportions, about three-fourths with the dominant character of tallness, and one-fourth with the character of dwarfness. Since dwarfness had seemingly disappeared in one generation, Mendel called it recessive, contrasting such traits with their dominant alternatives. In many similar experiments with the garden pea, Mendel obtained the same results with such characters as round or wrinkled seeds,

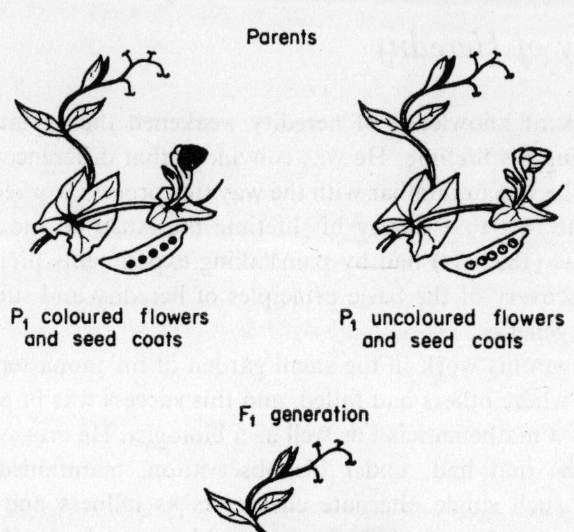

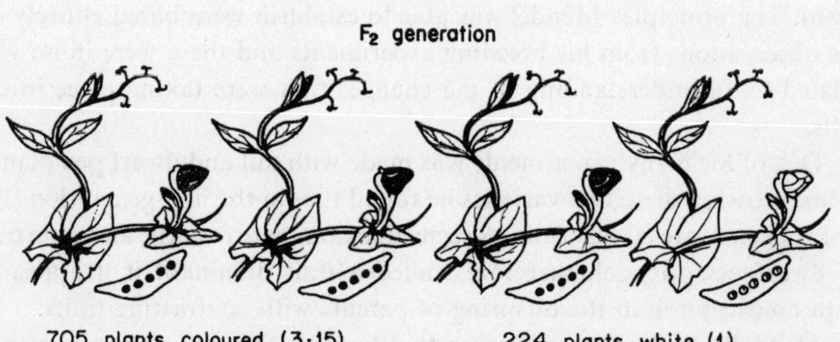

FIG. 39. Results of one of Mendel's earlier experiments when he crossed a pea plant with coloured flowers, coloured spots in leaf axils and coloured seed coats with the uncoloured variety.

coloured or white seed-coats, green or yellow seed-pods, and so on. Mendel did not use the word gene for genetic attributes such as dominance or recessiveness. Instead he spoke of hypothetical entities which he named units, factors and characters. Not until around 1910, well after Mendel's work was finished, was the name 'gene' coined.

Mendel rightly concluded that every one of his pea plants must surely have a pair of units for each trait. He arrived at this conclusion because some plants with a dominant trait, such as tallness, produced some offspring with the recessive trait, dwarfness. The parent plants, he reasoned, could hardly have been tall without having the dominant gene, and yet they could hardly produce offspring that were dwarfs without also possessing the recessive gene. Mendel thus assumed that if 'T' stands for tallness and 't' stands for dwarfness, the first generation offspring are a cross between tall and dwarf plants, and must contain one of each sort of unit 'Tt'.

Investigator	Yellow seeds		Green seeds		Total
	No.	%	No.	%	
Mendel 1865	6,022	(75·05)	2,001	(24·95)	8,023
Correns 1900	1,394	(75·47)	453	(24·53)	1,847
Tschermak 1900	3,580	(75·05)	1,190	(24·95)	4,770

FIG. 40. Summary of second generation (F_2) results in inheritance of seed colour in peas.

Naturally Mendel was only forced to assume a pair of units where the offspring produced were different from the parents. Thus plants that bred true for tallness, or dwarfness, might well have only contained one respective unit for each of these characters, but Mendel rightly assumed that even those plants which bred true also possessed pairs of attributes, though this time the pair consisted either of two dominant units for tallness TT, or two recessive units for smallness tt. Since Mendel's discovery the units now called genes of an organism have mostly been designated by such paired symbols, which for the trait being investigated indicate the organism's so-called genotype. Furthermore, if a pair of genes are identical they are called homozygous such as TT, two dominant genes for tallness, but if they are different (Tt) they are called heterozygous. Individual members of a gene pair, for example T tallness or t dwarfness, are known by the term allele. This Greek word, which means belonging to one another, refers only to individual members of a gene pair. Thus, if two gene pairs Tt and Cc are situated on a chromosome, T is an allele of small t and vice versa. T, however, is not an allele of C, both being members of a different pair. Finally, yet one more concept requires attention. We have seen that the genotype of an organism refers to its genetic make-up, or its gene formula Tt for a tall pea plant or tt for a dwarf plant. On the other hand, the outward appearance of an organism is referred to as its phenotype. The phenotype, therefore, expresses the outward appearance of a particular genotype.

Mendel's findings so far depended entirely on the results derived from one kind of garden pea. Later studies showed that the principles he had formulated were valid for almost all organisms. Before we can leave Mendel and his remarkable studies, however, we must briefly mention a number of important variations and complications. Firstly, there may be more than two different alleles for a particular trait. Because Mendel had worked only with the garden pea, he had not encountered this problem owing to the fact, as we have seen, that his alleles for each trait were limited to two. Thus his plants could only be tall or dwarf, their seeds wrinkled or smooth. There were no intermediaries. It is now clear that there are many traits in organisms represented by multiple alleles. Any two of these may occur together in the same cell in any combination, one each being contributed by the male and female parents respectively.

The human blood groups provide a good example to illustrate multiple alleles. Three main alleles may affect an individual's blood group, A, B and O. An individual may possess any two of these alleles but not more than two. The three alleles produce a number of interesting combinations. When alleles A and B are present in the same person, no dominance is shown by either. Both alleles are expressed in the phenotype, and the person is type AB. But alleles A and B are dominant to O, and only when both A and B are absent is type O expressed in the phenotype. There are thus four phenotypes associated with their respective genotype.

Phenotype	Genotype
Type A	AA or AO
Type B	BB or BO
Type AB	AB
Type O	OO

FIG. 41. Genetics of A-B-O blood types.

Although multiple alleles are responsible for some cases of varying traits, this phenomenon cannot completely explain the many characters in organisms that seem to vary over an often long and continuous range. Skin colour in human beings is an example.

To explain continuously varying characters such as human skin colour, human height, or for that matter the colour of the antirrhinums growing in your garden, the hypothesis has been developed that several different gene pairs may affect the same trait. Thus, a plant may have a number of different gene pairs that independently affect its colour. There is the added complication that each gene pair may possess dominant or recessive alleles. Consequently the number of different variants finally becomes very large. In 1908 the Swedish geneticist, Nilsson Ehne, showed that seed colour in wheat was inherited in this way. Now it is possible to explain, by means of hereditary principles, the reasons why parents produce such varying offspring.

Advances in genetics have shown the general invalidity of the concept of absolute dominance which Mendel found in his garden peas. Very often a dominant gene will appear incompletely in the heterozygous state and the offspring will be of an intermediate type. Today, because they are shown to be affected by age, temperature, sex hormones and other external factors, dominance and recessiveness of alleles is no longer considered a purely intrinsic property of genes.

Genes, Chromosomes and Evolution

IT WAS NOT until the turn of the century that K. Correns, H. de Vries and E. Tschermak, each working in a different European country, obtained results similar to those of Mendel, whose work had remained neglected for thirty-five years. Each of these three men acknowledged Mendel as the founder of the modern science of genetics upon which further hereditary studies could be built.

Although today there is abundant evidence for Mendel's hereditary theory, it is quite justifiable to ask, 'Are there really such things as genes and, if so, where are they, and what are they made from?' Answers to these questions were found when scientists studied the behaviour and structure of the chromosomes, those tiny sausage-shaped bodies which become visible in dividing cells. Research into those chromosomes involved in the production of gametes (the egg and sperm cells) revealed that the behaviour of chromosomes in cell divisions that produce gametes was parallel to the behaviour of the genes.

This relationship between Mendel's units of heredity and the chromosomes was already recognized by Walter S. Sutton (1876–1916), who published his results in 1902 and 1903.

When cells divide in the ordinary way (mitosis) the cell contents seem to be split roughly into two equal halves. The chromosomes, however, are always divided exactly during this process. At some time before the cell actually undergoes division, each chromosome forms a replica of itself. Thus, what originally was one chromosome before cell division, now becomes a pair of identical units. When a cell actually divides, one chromosome of each pair goes into each daughter cell. The daughter cells, therefore, have identical chromosomes which are replicas of the parent cell from which they were derived. With very few exceptions, individuals of the same species possess in their cells the same number of chromosomes. Thus the fruit-fly has 8 chromosomes, man has 46, a potato has 48, while some single-celled animals such as the amoeba may have as many as a hundred chromosomes.

Seeking to prove that genes were definitely physical units located in particular sites along the chromosomes, Sutton and his associates turned their attention to the biological link between generations of the vast majority of organisms, namely the tiny sperm and egg cells that unite during the process of fertilization to become the zygote.

Sutton argued that since genes are passed on from one generation to the next, they must be located somewhere within the sperm and the egg cells. Mendel had already shown, in his breeding experiments with the garden pea, that the male and the female gametes made equal genetic contributions to the organism which develops from their union. Sutton now insisted that if the male and female gametes both make equal contributions to the hereditary make-up of the offspring, the carriers of heredity (genes) ought also to be located in the same place in the egg and sperm.

Workers in this field had already known for some time that the sperm cell consists primarily of a nucleus, and although the egg cell (the female gamete) is often much larger in size than the sperm, its nucleus is similar to the nucleus of the sperm. Sutton concluded, therefore, that the nucleus was in fact the seat of heredity. But if this is so one major problem emerges. If all cells in an organism have the same number of chromosomes, and this also applied to the sperm and the egg, when the gametes fused during the process of fertilization, the resulting zygote would contain double the number of chromosomes! In man this would mean that the chromosome number would be doubled from 46 to 92, and if the process continued in

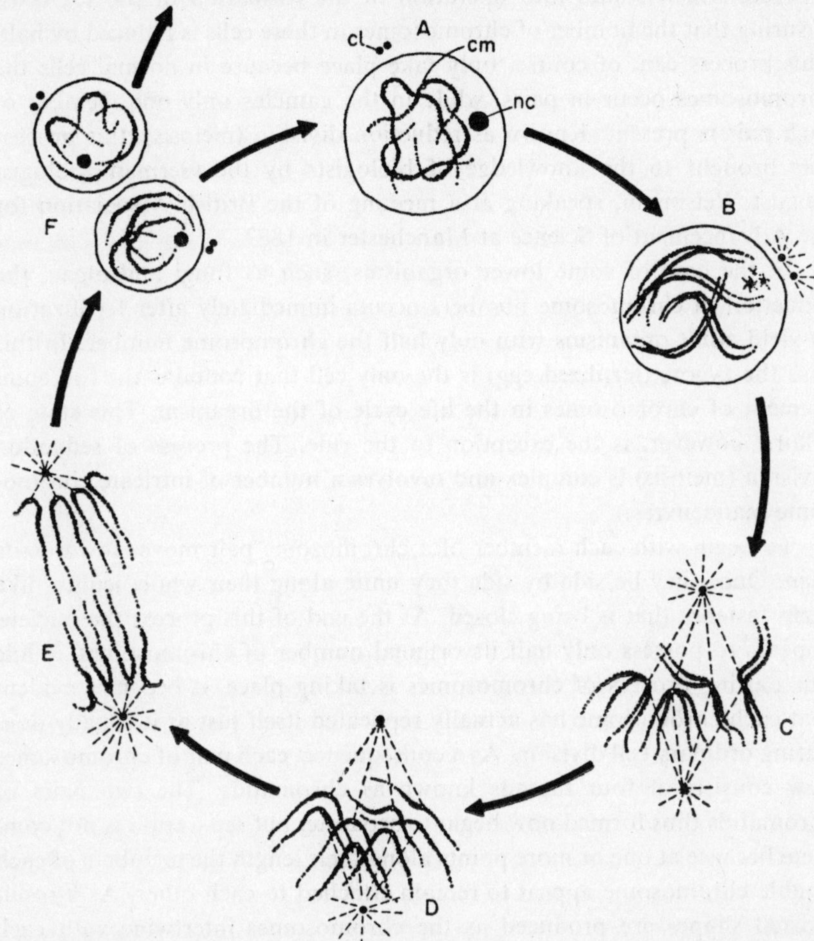

FIG. 42. Diagram of cell division by mitosis illustrated by a cell with two pairs of chromosomes. (A) 'Resting' nucleus: cl, centrioles; cm, centromere; nc, nucleolus. (B) Each chromosome is visibly double, the duplicates still joined by the centromeres, which have not yet divided. (C) The double chromosomes are aligned at the equator between the poles of the spindle. (D) The centromeres have split and the chromosomes are beginning to move toward the poles. (E) The chromosome halves have moved to the poles, one complete set (two pairs) going to each pole. (F) Each chromosome set is included within a new nucleus, and two new cells have been formed.

subsequent generations the cells would in due time contain many thousands of chromosomes.

Through the course of evolution a special safety mechanism has been evolved, which comes into operation in the formation of the sex cells, ensuring that the number of chromosomes in these cells is reduced by half. This process can, of course, only take place because in normal cells the chromosomes occur in pairs, while in the gametes only one member of each pair is present. Known as reduction division (meiosis), this process was brought to the knowledge of biologists by the German zoologist August Weismann, speaking at a meeting of the British Association for the Advancement of Science at Manchester in 1887.

In the case of some lower organisms, such as fungi and algae, the reduction in chromosome numbers occurs immediately after fertilization to yield adult organisms with only half the chromosome number. In this case the zygote (fertilized egg) is the only cell that contains the full complement of chromosomes in the life cycle of the organism. This state of affairs, however, is the exception to the rule. The process of reduction division (meiosis) is complex and involves a number of intricate chromosome manoeuvres.

To begin with each member of a chromosome pair moves towards its twin. Once they lie side by side they unite along their whole length, like a zip fastener that is being closed. At the end of this process the nucleus appears to possess only half its original number of chromosomes. While this pairing process of chromosomes is taking place, it becomes evident that each chromosome has actually replicated itself just as it usually does during ordinary cell division. As a consequence, each pair of chromosomes now consists of four threads known as chromatids. The two pairs of chromatids thus formed now begin to separate, but separation is not complete because at one or more points along their length the members of each double chromosome appear to remain attached to each other. As a result twisted shapes are produced as the chromosomes intertwine with each other.

It was found that during this process of 'crossing over' some exchange of chromosome material between the pairs actually takes place. Then, the double chromosomes begin to move slowly away from each other to opposite ends of the cell where two daughter cells each containing double chromosomes are formed. The number of double chromosomes in each daughter cell is exactly half the normal chromosome number of the organism. These new cells now undergo another division, but this time with no

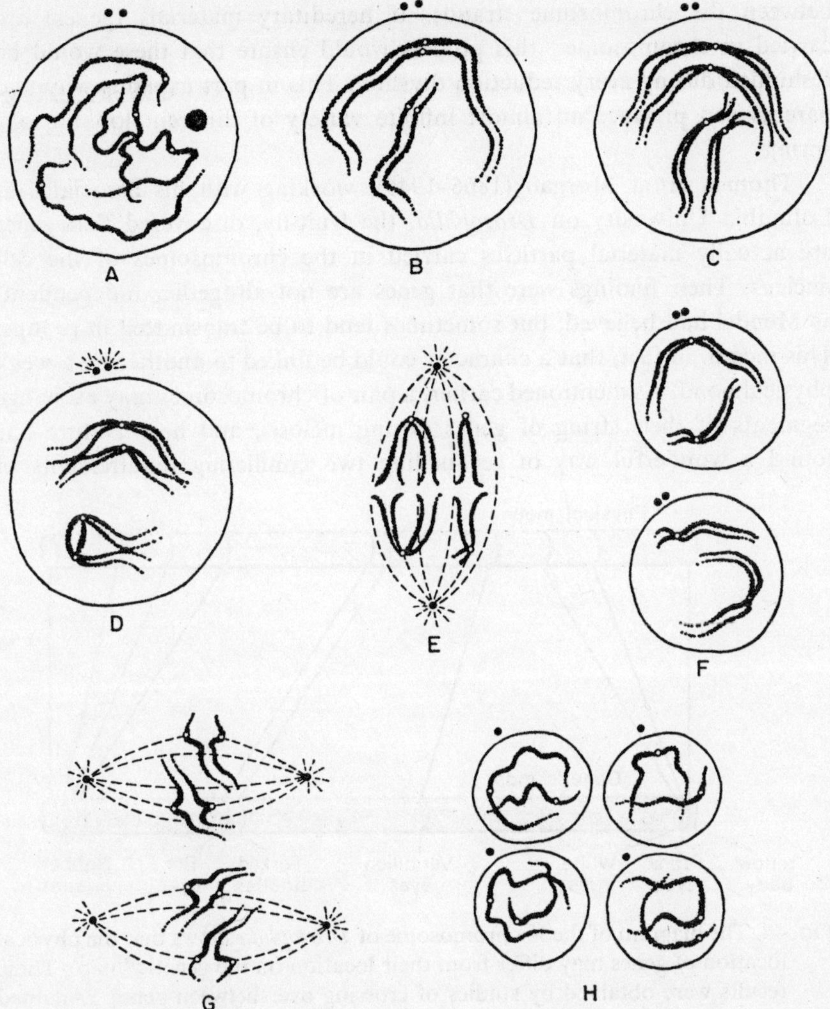

FIG. 43. The progress of meiosis or reduction division during the formation of the sex cells (gametes). A diagrammatic representation illustrated by two pairs of chromosomes.

replication of chromosome strands. Instead, the two chromatids of each chromosome separate, one to each daughter cell.

The result of the two divisions in meiosis is four cells each with a single member of each pair of chromosomes. These, then, are the egg and sperm cells. The elaborate and complex process of meiosis involves, as mentioned earlier, a number of manoeuvres which ensure an exchange of material between the chromosome strands. If hereditary materials (genes) are carried in chromosomes, this process would ensure that these would be reshuffled during every reduction division. This in part explains why two parents can produce an almost infinite variety of different looking off-spring.

Thomas Hunt Morgan (1866–1945), working with his associates at Columbia University on *Drosophila*, the fruit-fly, discovered that genes are actually material particles carried in the chromosomes of the cell nucleus. Their findings were that genes are not altogether independent, as Mendel had believed, but sometimes tend to be transmitted in groups. This meant, in fact, that a character could be linked to another by a weak physical bond. As mentioned earlier, a pair of chromosomes may exchange segments of their string of genes during meiosis, and here Nature has found a wonderful way of reconciling two conflicting requirements of

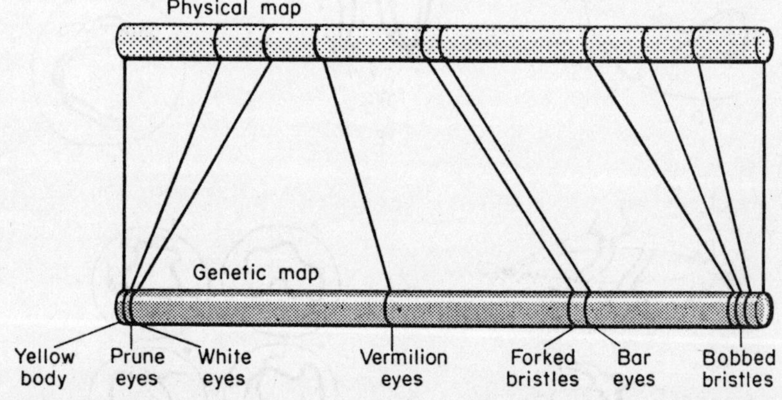

FIG. 44. The diagram of the X-chromosome of *Drosophila* shows that the physical location of genes may differ from their location on the genetic 'map'. These results were obtained by studies of crossing over between genes. (Adapted from Biological Sciences Curriculum Study, ***Biological Science: Molecules to Man***, Houghton Mifflin Co.)

inheritance and evolution. On the one hand, total disorganization of the genes in a cell would make the reproduction of cells exceedingly difficult when it is realized that the fruit-fly has something of the order of five to ten thousand individual genes. If all these genes were lying loose in the cell nucleus like a heap of marbles, the problem of passing them on in exactly equal numbers to every daughter cell would be extremely difficult to solve. The task is reduced to more manageable proportions by the fact that genes are apparently grouped on the chromosomes, which in the case of the fruit-fly number eight (four pairs). Thus the cell has to cope with only eight objects instead of a possible ten thousand. On the other hand, if genes were for ever fixed in one and the same chromosome, the cell would lack the flexibility for recombination of genes which is so essential to the evolutionary process. The problem is solved by the ingenious arrangement whereby genes may change their place of domicile during the formation of the sex cells, crossing over from one chromosome to its partner.

How Sex is Determined

MAN HAS PUZZLED for centuries over the fact that most creatures on our planet reproduce sexually. Why has the experiment of evolution attached such an importance to this aspect of living matter? Mendel's theory showed that the sexual methods of reproduction provide an elaborate pool which serves the function of recombining genes in new ways, thus permitting living things to experiment with a practically limitless range of possible variations. The number of different gene combinations in a species is very large, and provides a vast reservoir of potential variability upon which the species can draw for its evolutionary experimentations while still maintaining its fundamental characteristics. This is true of all species that use the method of sexual reproduction, be it a bacterium or man.

Let us now return for a moment to the laboratory of T. H. Morgan, where thousands of fruit-flies were being raised in bottles and fed on mashed bananas. In nature these flies possess deep red eyes, but during his breeding experiments Morgan encountered a male fly with white eyes. When he mated this white-eyed male with a red-eyed female, he found that in the first generation the offspring consisted entirely of red-eyed flies. Morgan thus rightly assumed that the gene for red eyes was dominant and the gene for white eyes was recessive. He then allowed this first generation to mate among themselves, and found that their offspring was in the ratio three-quarters red-eyed flies to one-quarter white-eyed flies. This again was not an unexpected result, although Morgan discovered that all the white-eyed flies were males! White eyes, therefore, seemed in some inexplicable manner to be linked to the sex of the individuals.

This result indicated to the scientists the existence of a fundamental difference in the chromosomes of the male and female fruit-flies. Further researches on the cells of *Drosophila*, the fruit-fly, subsequently confirmed this. In either sex three of the four pairs of chromosomes appear similar under the microscope, but the fourth pair of chromosomes shows marked differences. Females have a pair of straight so-called X-chromosomes, while males possess one straight X-chromosome and one hooked Y-chromosome, the pairs being described by the symbols XX and XY respectively.

The discovery of two kinds of chromosomes in *Drosophila* suggested that they might provide an explanation for the actual determination of sex. This, indeed, is precisely what scientists working with fruit-flies later discovered. The sex of a fly depends on whether the egg is fertilized with a sperm bearing an X-chromosome or one with a Y-chromosome; the former combination resulting in a female XX, the latter in a male XY. It has since been discovered that human cells show a similar pattern of sex chromosomes to those of the fruit-fly.

Organism	Male	Female
Drosophila	XY	XX
Mammals	XY	XX
Birds and butterflies	XX	X–
Grasshoppers and roaches	X–	XX

FIG. 45. Sex chromosome patterns in a number of organisms.

In most groups of animals, and in the majority of those higher plants that have separate male and female individuals, it is the male sex that produces two kinds of sperms or pollen grains. But in certain fishes, amphibians, all birds, and many insects, the situation is reversed so that there are two kinds of eggs and only one kind of sperm.

Morgan's experiments with the fruit-fly reveal that not only do sex chromosomes determine the sex of an individual, but clearly carry genes for other hereditary traits. While the members of a pair of ordinary chromosomes (also known as autosomes) in a cell have similar genes, sex chromosomes differ in appearance and occasionally carry different genes. The inheritance of white eyes in the fruit-fly can be explained most simply by the theory that the gene for white eyes is present only in the X-chromosome, and that the Y-chromosome has no corresponding allele. It is assumed to be quite inactive in the inheritance of white eyes. If this theory is true, then even a single recessive gene in the X-chromosome could become recognizable in the male fly providing no corresponding dominant allele on the Y-chromosome masked its effect.

Morgan thus demonstrated that there was a direct relationship between a particular heritable character and a specific chromosome (in this case white eyes and the X-chromosome respectively). This provided perfect evidence in confirmation of Sutton's theory that genes are located in chromosomes. Since man, as mentioned, possesses a pair of sex chromosomes similar to that of the fruit-fly, it is reasonable to assume that some human hereditary traits might well be sex-linked, too. A good example is a form of colour-blindness—the inability to distinguish between red and green. It has always been known that there are many more colour-blind men than women. The reason for this is not difficult to understand. If a man inherits an X-chromosome from his mother that carries the genes for colour-blindness, he will be colour-blind. In order for a woman to be colour-blind she must have an X-chromosome from her mother and an X-chromosome from her father, both carrying the gene for colour-blindness. If, for example, one X-chromosome in ten carries the gene for colour-blindness, one male in ten will be colour-blind, but the probability of a female having two X-chromosomes each carrying the colour-blind gene is much smaller; hence there are fewer colour-blind women than colour-blind men.

A work published by Calvin B. Bridges (1889–1938) finally clinches the fact that chromosomes are indeed the carriers of hereditary material. Bridges was one of Morgan's graduate students at Columbia University, and at the time was studying how a gene which produced especially bright red eyes in the fruit-fly, known as vermilion eyes, was transmitted in the fruit-fly. The gene in question was found to be recessive and like the gene for white eyes appeared to be associated with the X-chromosome. From what we have just heard of sex-linked inheritance, it seems clear that vermilion-eyed females crossed with normal males should give rise to females that appear normal and to vermilion-eyed males. This, indeed, is what Morgan found in the vast majority of cases, but a vermilion-eyed female did appear very occasionally—once in every two thousand flies. Nevertheless, this peculiar occurrence did raise some doubts about the validity of the chromosome theory of heredity.

Bridges reasoned that, according to this theory, a vermilion-eyed female could not be produced unless there were two recessive alleles for vermilion eyes present, because the vermilion eye colour is recessive to the red eye colour of the natural flies. One of the two alleles required was contributed by the mother in the egg cell. The question researchers had to ask themselves was, 'Where did the other recessive allele come from?' Bridges found the right answer and put forward the brilliant theory that the two X-chromosomes which should have separated during the formation of the egg did not in fact do so, both being transmitted as a pair to the daughter together with the usual single male Y-chromosome from the father fly.

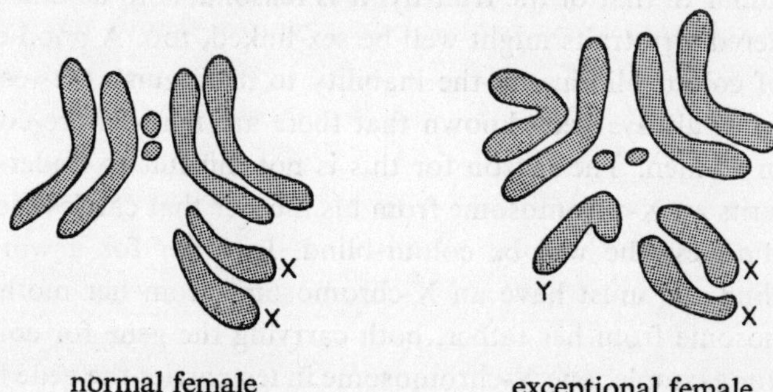

FIG. 46. The chromosomes from a normal *Drosophila* female with two X-chromosomes (left), and from an exceptional vermilion-eyed female with two X-chromosomes and one Y-chromosome. (Adapted from Biological Sciences Curriculum Study, *Biological Science: Molecules to Man*, Houghton Mifflin Co.)

The convincing test for this hypothesis came when Bridges and his team examined the cells of these exceptional flies under the microscope. Bridges found that his theory had been correct. There were, indeed, three sex chromosomes instead of the normal two, one being a Y-chromosome derived from the father fly, the other being two X-chromosomes from the mother fly. The entire series of experiments was a dramatic and convincing demonstration that genes were actually located on the chromosomes. This decisive breakthrough finally established the chromosome theory of heredity.

The Material of the Genes

THE EVOLUTION of life is intimately connected with the principles of heredity and the concept of natural selection. The genetic basis of evolution, however, must be elucidated if we are to comprehend this truly remarkable process. What are genes made of? How do they function? What changes in them may affect the course of evolution?

We have in part familiarized ourselves in the previous chapter with the basic unit of life (the cell), and with some of its internal structures which make possible the self-replication of organisms. The nucleic acids, so brilliantly elucidated by Crick and Watson in 1953, at last gave the scientific world a plausible explanation of self-replicating molecules which could form the basis of heredity. Usually, the cell and its nucleus can be seen under the microscope, but the chromosomes only become visible at certain stages in the life of the cell, when it divides to give rise to two daughter cells. During this process of cell division the chromosomes appear as rod-like structures which, in thin tissue slices, can be stained with suitable dyes to make them stand out clearly against the rest of the cell.

Over the last thirty years, no creature has helped scientists to elucidate the structure of the chromosomes more than the tiny fruit-fly, *Drosophila*, and its relatives. For reasons as yet unknown, the chromosomes in the salivary gland cells of the larvae of this fly remain in an elongated condition, and proceed to divide over and over again with daughter strands remaining in close juxtaposition. Thus they become easily visible as long structures. When the larva of such a fly is ready to enter the pupal stage in preparation for transformation into the winged adult, the total length of the chromosomal strands in a single salivary gland cell may be as long as one millimetre. If examined under a powerful microscope, a number of detailed structures can be seen on the chromosomes. Although it is not possible to detect the individual genes under the microscope, these chromosomes are marked throughout their length by an intricate series of wide, narrow, and dotted bands.* These indicate the position of genes. So clear are these markings that it has been possible to map each chromosome throughout its entire length.

* Drs O. L. Miller and B. R. Beatty of the Oak Ridge National Laboratory have produced electron micrographs in which are visible genes in the process of manufacturing ribosomal RNA (*Science*, 164, 3882, 955).

Fruit-flies are estimated to possess between five and ten thousand different individual genes. These are responsible for such characters as eye colour, length and shape of wings, body size, various aspects of behaviour, and so on. Normally, a full complement of genes is present in every cell of the body. They govern the functions of the cell and are primarily responsible for its properties and behaviour.

After the discovery of DNA, a primary concern of scientists was to prove that this remarkable substance was the actual material of the gene, and therefore the basis of heredity.

Torbjörn Caspersson, the Swedish biologist, examined cells under the influence of ultra-violet light. It had been known for some time that the nucleic acids DNA and RNA strongly absorbed this type of radiation. Aided by photoelectric cells and other complex equipment, Caspersson was able to measure the absorption of ultra-violet light by the nucleic acid. This enabled him to calculate the actual amount of DNA present in the cells. He showed that, whenever cells are in the process of rapid growth, these acids are present in larger quantities than normally. Scientists deduced, therefore, that nucleic acids probably played an important part in the formation of new cellular substances.

About twenty years later, scientists in France and the United States began experiments to compare the relative amounts of the nucleic acid DNA in different types of cell. They found that, with the exception of the gametes, the cells of the bodies of animals belonging to the same species contained identical quantities of DNA.

Particularly significant were their findings that the reproductive cells contained only half these amounts of DNA. Since the reproductive cells carry only half the number of chromosomes of the ordinary cell, the result strongly hinted that DNA was closely connected with the hereditary factors in chromosomes.

The clearest evidence that DNA is the hereditary material, and that genes are made from DNA, came from experiments with bacteria. Bacteria belong to those groups of cells that, although they seem to lack a distinct nucleus, have in common with all living organisms either one or both the nucleic acids DNA and RNA within their cells.

In the late 1920s the British doctor, Frederick Griffith, was studying an organism which causes pneumonia, *Diplococcus pneumoniae*, when he made the discovery that these bacteria occurred in two different types. In one type, each pair of bacterial cells was surrounded by a capsule of a sugar-like substance. The other type had no capsule around the cells. The difference between them is that the bacteria with capsules are highly dangerous, and those without capsules are harmless. Griffith killed the virulent bacteria by heating them to 60°C., and, as expected, when injecting them into mice he found them to be harmless. In the second experiment he injected a mixture of heat-killed dangerous bacteria, together with live harmless bacteria, and was surprised to find that the mice sometimes mysteriously died of pneumonia. Even more surprising was the fact that the blood of the dead animals was invariably swarming with live virulent bacteria which had somehow acquired capsules.

At that time, Griffith was unable to find a satisfactory explanation for this remarkable transformation. Later workers, notably Martin H. Dawson of the Rockefeller Institute, found that he could bring about the change by simply mixing the dangerous heat-killed bacteria with the live harmless ones in a nutrient solution in a test tube. Observations proved that, after some time, he had obtained dangerous living bacteria surrounded by their tell-tale capsules. These experiments made it clear that the agent responsible for this remarkable transformation must form a part of the bacterial cell. It might well be an organic substance capable of altering the genetic constitution of this particular bacterium.

The pieces of the jig-saw of heredity were beginning to fall into place, but the material responsible had still to be isolated and chemically analysed.

This task fell to Dr O. Avery and his young team of research physicians working at the Rockefeller Institute in New York. They undertook to identify chemically the substance that was able to change a perfectly harmless bacterium into a highly dangerous one. The technique used was based on a process of elimination. By means of enzymes, they selectively broke down the known chemical components of the bacterial cell. First they dissolved away the sugar-like capsule, and thus stripped the bacterium of its covering. They found that the bacteria which had their capsules removed still retained their dangerous potential and were able to transform harmless bacteria into virulent ones. Obviously it was not the bacterial coating which carried the dangerous characters.

They then set about to remove the protein portion of the bacterium by using a specific enzyme that attacked proteins. Once again, results were negative. The mysterious transforming agent still remained. After eliminating a number of other known substances, the experimenters were finally left with a substance that consisted almost entirely of the two nucleic acids DNA and RNA. The scientists now set about to purify the transforming agent further, and then subjected it to a detailed chemical analysis. When they compared the chemical composition of the transforming agent with that of DNA, the researchers were delighted to find that the two were almost identical.

In their final experiment they mixed the transforming agent with an enzyme known to destroy DNA. The results were positive. The transforming agent had lost its ability to alter the genetic constitution of the harmless bacteria. The culprit had at last been tracked down. It was the nucleic acid in the dangerous bacterial cells that had brought about the permanent hereditary change in the harmless bacteria.

When their findings were published in 1944, it became clear that Avery and his associates had discovered the genetic material of the cell—the same substance Friederich Miecher had investigated seventy-five years earlier.

Two Important Functions of the Gene

IT WAS SOON REALIZED that genes have two major roles to play. Firstly, there is the process of DNA replication which enables all cells descended from the fertilized egg to receive replicas of its DNA molecules.

Secondly, by their action genes control every step of an organism's development, right from the moment of fertilization to the full functioning of the adult organism. This is confirmed when we observe how closely offspring resemble their parents in structure, function and behaviour.

An interesting example how genes function on the cellular level comes from the researches of Vernon Ingram and his team of scientists working at Cambridge University. For a long time doctors had been aware of the fact that a certain blood defect in man, sickle-cell anaemia, was inherited as a simple recessive trait. In this condition, the red cells become sickle-shaped when the oxygen concentration is low.

Scientists now carefully analysed the chemical structure of the oxygen-carrying pigment haemoglobin, which is a protein and imparts the red colour to the blood cells. They compared the chemical structure of haemoglobin in normal individuals to that found in individuals afflicted with the sickle-cell trait. Their results were clear and revealing. When they analysed the haemoglobin molecule they found it consisted of 300 amino acids of 19 different kinds, all arranged in a precise 3-dimensional structure. Step by step they broke down the molecule until they could detect the difference in amino acid units between the normal haemoglobin and the sickle-cell haemoglobin. Ingram was now able to show that sickle-cell haemoglobin differed only in one amino acid from normal haemoglobin. At one particular site in the structure glutamic acid was replaced by valine. Valine, an essential amino acid in many proteins including haemoglobin, was disadvantageous at this particular point in the structure of the haemoglobin molecule.

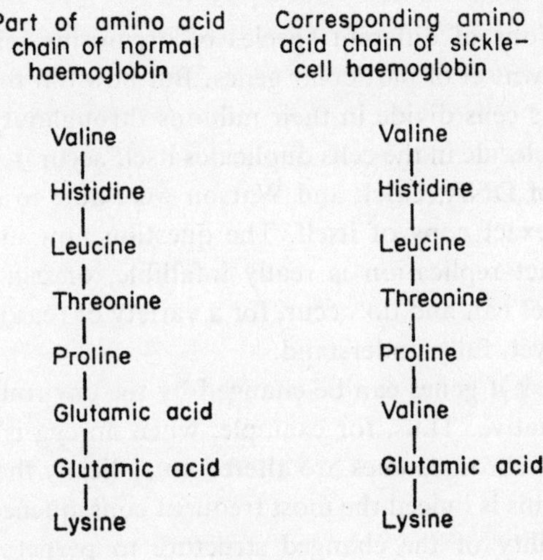

Fig. 47

The evidence clearly showed that the slightly altered gene responsible for the sickle-cell condition of the red blood cells causes a valine unit to be substituted for a glutamic acid unit at one particular site in the haemoglobin molecule. The interesting fact is that only one amino acid in a single situation involving 300 units is affected by the altered gene. From this study, two points emerge. Firstly, a change in a gene can alter a single unit of a very large molecule such as, for example, a protein, without altering the rest of the molecule. Secondly, a slight chemical change such as the substitution of a single amino acid unit in the normal haemoglobin molecule can profoundly alter the red blood cells of a person who carries this hereditary trait.

Gene Mutations and Evolution

THE OBSERVATIONS and experiments described have established genes to be the carriers of heredity, and also support the view that genes consist largely of nucleic acids, usually DNA. Specific nucleic acid molecules control particular chemical reactions in the cell. It follows that the life activities of the cell are largely controlled by the combined action of the genes.

Many millions of different species of organisms inhabit our planet, each with its own set of particular genes. But how did this huge variety of genes arise? As cells divide in their millions throughout the living world, every DNA molecule in the cells duplicates itself accurately. By elucidating the structure of DNA, Crick and Watson were able to show how a gene can make an exact copy of itself. The question now arises whether this process of exact replication is really infallible, or can mistakes occur? Clearly mistakes can and do occur, for a variety of reasons most of which we do not, as yet, fully understand.

When we ask if genes can be changed by the environment, the answer is in the affirmative. Thus, for example, when an egg is boiled the genes contained in its chromosomes are altered so radically that they will never divide again. This is indeed the most frequent consequence of induced gene changes—inability of the changed structure to perpetuate itself by cell replication. A gene may also be altered without losing the ability to replicate itself. The altered structure is then copied when reproduction occurs. But if the gene is changed so is its genetic message, and a new genetic message will in all probability cause a slightly different protein to be synthesized in the cell, as the example of the sickle-cell gene showed. Changes that affect the genetic message in a cell are called mutations.

Let us at once state that only those mutations which directly affect the sex cells can have any direct bearing on the offspring. The types of mutations that can occur are numerous, and range from very minor and undetectable effects to mutations that will prove immediately lethal to an organism. If, for example, a mutation in a blackbird produces a very slightly longer and more pointed beak, this mutation would not be immediately noticeable, but its long-term effect might well be to help the affected individuals to become more proficient insect catchers. Thus, in an environment where insect food was scarce, this bird would find itself with a slight advantage over those birds where the mutation did not occur. In another hypothetical example, a mutation may adversely affect the secretion of digestive juices in the stomach. Such a mutation would probably prevent the bird from obtaining enough nourishment, with the likely result that it would soon die.

Lethal mutations are also known in human beings. Thus haemophilia is an abnormality found in humans caused by a recessive sex-linked gene. The blood of a person who suffers from this disease fails to clot on exposure to air, and even a slight wound may have fatal results. New mutations to haemophilia are rare but are known to have occurred in certain families.

In any organism some genes mutate rather more frequently than others. Some may mutate as frequently as one in every five or six thousand sex cells. Others may mutate once in every five or six million sex cells. In general, however, mutations do not occur very frequently in any one gene, but if we consider that man has an estimated number of 10,000 genes and if each of these mutates at the rate of 1 in 100,000 mathematically a mutation of some sort or another is likely to occur in one out of every ten sex cells. To date there are about 3,000 million human beings living on our planet. The number of new mutations in each generation, therefore, is truly astronomical. This, of course, applies equally to all other organisms, and consequently mutations provide a rich storehouse for the evolutionary process.

As mentioned earlier, the causes of most mutations in nature are unknown, but over many years past scientists have perfected techniques to increase in laboratory animals the number of mutations by artificial means. The most direct course is by treating the reproductive cells with various external agents. These agents roughly fall into three categories. It has been shown that the mutation rate increases if the temperature of the reproductive cells is artificially increased within a narrow range without killing the cells.

Similarly, when cells are treated with certain chemical compounds (mustard gas is an example) similar affects can be obtained. Thirdly, high energy forms of radiation, such as X-rays, beta-rays, gamma-rays, neutrons and others, are the best known and studied causes of mutations.

Radiation experiments have produced some striking increases in mutation rates. H. J. Muller, working at the University of Texas in 1927, was the first man to apply this method in practice. As a graduate student of Columbia University, Muller had worked with T. H. Morgan and C. B. Bridges on some of the earlier genetic experiments with fruit-flies. Depending on the amount of X-rays applied, Muller found that mutations may be tens or even hundreds of times more frequent in the offspring of irradiated flies than in the non-irradiated control flies. There was also no doubt that the X-rays increased the rate of occurrence of lethal mutations.

High energy forms of radiation have also produced mutations in many experimental animals and there is every reason to believe that they are mutagenic in all organisms, including man. Of particular importance is the evidence that the number of mutations induced is proportional to the amount of X-rays reaching the sex cells.

Much discussion has taken place in recent years concerning the increased use of X-rays in medicine, and wise steps have been taken to confine their use to the diagnosis of disease, where they form a most valuable tool. Regrettably, most organisms are becoming increasingly exposed to the radiations produced by the release of atomic energy, and unless this danger is brought under strict control it will cause a serious threat to future generations of all forms of life.

How Flies and Bacteria Can Change

EVOLUTION is the process by which nature experiments with new varieties of life which arise from existing ones.

Flowering plants have evolved from ferns and mosses, and birds and mammals from reptiles, and reptiles from amphibians, and amphibians from fishes. In general, the trial and error methods of evolution are extremely difficult to observe because the human life span is too brief to watch evolution at work except in those cases where many generations of an organism succeed each other in rapid succession, as is the case with many insects and micro-organisms such as the bacteria or viruses.

Many examples of evolution by selection in response to poisons which kill insect pests are known. The particularly powerful insecticide dichloro-diphenyl-trichloroethane, known as DDT, when first used, killed nearly all house-flies. Yet in 1947, a number of years after it had come into general use, reports were received from Italy and Sweden that house-flies had become relatively resistant to DDT. During the first year or two of DDT use, when hordes of flies were killed by the insecticide, there were a few flies that were able to resist the drug because of their genetic constitution. These surviving flies then reproduced more of their own type in the area where this insecticide had been in use. Thus, in the following years, DDT-resistant house-flies were able to breed more successfully, making DDT less effective. Hence, wherever the drug was used indiscriminately over a long period of time, the drug-resistant flies replaced the normal type. DDT does not directly affect the genes; but, by killing those flies that lack genetic resistance to the insecticide, it simply picks out the flies whose genes, by accident, have been so constructed that they can resist the drug.

Many other examples can be quoted, but particularly interesting and important are those cases where certain bacteria have developed a resistance to penicillin and other antibiotics. Such an example is the bacterium *Staphylococcus aureus*. In one experimental study, about 100 million bacterial cells were placed on a nutrient medium containing a weak dose of penicillin. Less than ten bacterial cells survived, but the offspring of these cells multiplied well on the medium containing the weak penicillin dose. When this dose was doubled, most of the bacteria succumbed. The few relatively resistant survivors were isolated and exposed to still higher concentrations of penicillin. By repeating this process five times, bacteria were obtained which could grow in the presence of a penicillin dose two hundred and fifty times as strong as that which killed most of the bacterial cells in the original culture.

Again we see the work of selection. As the environment changes, in this case as more penicillin is added, one type of bacterium gains an advantage over the other. Fortunately, the majority of bacteria which have become resistant to antibiotics have also become dependent on the drugs, so that they are unable to grow satisfactorily without them. This means that, in the absence of the drugs, normal bacteria vulnerable to antibiotics will gain the upper hand again in the struggle for existence. They will then eliminate the drug-resistant bacteria. Here we have one reason why drug-resistant strains are often found in hospitals, where they are able to thrive in an environment which provides them with the necessary antibiotics on which they have become dependent.

How Moths Adapt to their Environment

ONE OF THE BEST KNOWN EXAMPLES of observed evolutionary changes concerns the development of dark, melanic varieties in several species of moths. These changes were confined to the industrialized regions of Western Europe, particularly in the British Isles. Each of these moth species has two varieties, one of light colouration and the other dark. Experiments have shown that the two varieties usually differ in one pair of genes, the allele for dark colouration being dominant to the light. Yet at one time the light variety was commonly found in the countryside while the dark ones were rare or altogether absent. From the middle of the nineteenth century naturalists began to observe that the dark variety of moth was occurring more often, particularly in the vicinity of large industrial towns. This development has continued in many industrial regions until today the moth populations in some areas consist mostly of dark species with or without an admixture of light ones.

Scientists believe that one of the reasons for this change was that the dark varieties were generally more vigorous than the light ones. The entomologist E. B. Ford was actually able to demonstrate the superior vigour of the dark variety in one particular species. The question now posed was why if the dark forms were more vigorous did they replace the normal light ones only in industrial regions? The answer is that light forms compensate for their lack of vigour by being protectively coloured. They are, in fact, beautifully camouflaged by matching the colouration of their surroundings, and thus they have a better chance of remaining undetected by birds than the more conspicuous dark varieties. But in industrial regions where the surrounding countryside is often contaminated with soot the situation alters. Here on the often blackened tree trunks the light-

FIG. 48. *Left* Two peppered moths on a tree with light-coloured lichen. *Right* Two peppered moths on a tree with dark-coloured bark.

coloured moths are no longer camouflaged, while their darker relatives now melt into the background of their surroundings and are thus overlooked by their enemies. The changed environment of an industrialized region removes the disadvantages encountered by the dark varieties and so they spread and eventually replace the light-coloured moths.

Darwin's Finches may have Evolved from One Species

THE FOURTEEN SPECIES of finches which inhabit more than a dozen islands of the Galapagos group have attracted the interest of a number of naturalists and much evidence has been accumulated to support the theory that all these finches may have evolved from one mainland species.

Let us now speculate how this may have come about and how evolution could have affected this remarkable group of birds on these isolated islands.

Six species of finches live on the ground, and these inhabit the arid coastal regions. Another six species live in the canopies of the island's trees and feed almost exclusively on insects, which live in the humid forests. Then there is a warbler-like finch which feeds on small insects near the ground in both the dry and humid regions of the islands. Lastly, there is an isolated species living in the more distant Cocos Island.

Of the ground finches, three species eat seed, and differ from each other mostly in the size and shape of their beaks, which are adapted to different sizes of seeds. The prickly pear cactus provides food for a fourth ground species with a much longer and pointed beak. These four ground species are found living together on most of the islands. The remaining two species of ground finches enjoy a mixed diet and seem to be confined to the outlying islands.

The tree finches, too, while on the whole similar to each other, differ chiefly in body size and in the size of their beaks. The size of beak apparently depends on the size of insects the birds feed on. A fourth species of tree finch is a vegetarian with a parrot-like beak suited to a diet of buds and fruits. The mangrove swamps are occupied by a fifth insect-eating species, but the most interesting finch of them all belongs to a sixth species that behaves like a woodpecker yet resembles a finch. In the true fashion of a woodpecker it climbs tree trunks in search of insects, using its chisel-like beak to excavate them from the bark. But while it has the beak of a woodpecker its tongue is too short and therefore it has to adapt itself to a feeding method unique in the bird world. This finch is usually seen carrying in its beak a cactus spine or small twig which it uses as a probe to dislodge insects from their hiding-places. When an insect emerges the bird quickly drops its stick and devours the insect. Tool-using seems largely to have been confined to man and the apes, and it is indeed remarkable that a bird follows such an advanced behaviour pattern. Recent studies in animal behaviour seem to indicate that birds are a good deal more intelligent than was originally thought.

Darwin's finches are so similar to each other in most features excepting their beaks that the evidence suggests they have all evolved from one original colonizing form which came to these islands from the South American mainland many thousands of years ago. Here, then, might be circumstantial evidence for the origin of new species by means of geographical isolation. The most unusual aspect about these finches is that several species today live on the same island.

Let us now go back in time to a distant period when there were as yet no finches to be found on the Galapagos Islands. One day, during a gale in the coastal regions of the South American mainland, a few birds from one particular species were blown over the sea to settle on two of the islands. Whilst one island was well forested and had an abundance of insect life, the other island was rich in seed-producing plants. If the original finches that came from the South American mainland were seed-eaters, those finches that landed on the island rich in insect food may gradually have evolved features which suited them best for an insect diet—for example a sharp pointed beak. But how was it that a mainland finch which chiefly fed on seeds gradually evolved into an insect-eating bird? This question can be answered by Darwin's theory of natural selection. The arrivals on an island well stocked with insects but relatively poorly supplied with seed-growing plants would face a situation which allowed them to breed comfortably for a number of generations until their numbers became so large that the seed food on the island became inadequate, and famine conditions resulted. Probably during the time of plenty, when all the finches could find sufficient seeds for their needs, a few might develop slight modifications of beak structure rendering them more useful in picking up insects to supplement their diet. Then, when seeds eventually began to run short the finches that had previously adapted themselves to insect food would probably find it easy to compete with their starving brothers for mates and breeding territories. The end result of all this might well have been that the insect-eaters rapidly increased in number at the expense of the seed-eaters. A few thousand years later, a visitor might find that the old species of seed-eaters which had originally colonized the island had completely disappeared to be replaced by a similar finch with a pointed beak and insect-eating habits. The change had taken place gradually by the process of natural selection.

At first there may have been a few individuals where a gene mutation supplied them with beaks that were at variance from the majority by being slightly more pointed and thus more suitable for catching insects. These birds were more successful in their quest for insect food, and thus when seed famine threatened, these finches gained an advantage over their relatives. We know that changes in an animal that take place on the genetic level are inherited from generation to generation, and thus the birds with more suitable beaks for insect-eating would pass on this advantage to succeeding generations.

Advantageous genetic variations in a species are exceedingly rare, but they do occur and form the basis for evolutionary changes. Over a long period of time, random genetic changes occur in all forms of life. Those that are advantageous to a particular creature will assist towards improving its position in its particular environment. For example, a finch living in an environment where insect food is plentiful will have a more pointed beak. On the other hand, detrimental changes on the genetic level will usually not be retained, because the afflicted individuals will find it difficult to compete with their normal relatives, and thus are likely to die out.

Among the finches on this particular island genetic variations leading to more pointed beaks were obviously advantageous to the species, and thus this trait continued to develop whenever a random genetic variation gave additional advantages to the insect-eating birds. On the second of the two islands, conditions may have been different, favouring the original mainland species with short stout beaks suitable for cracking seeds. Here, selection would favour those birds where genetic changes would lead to advantages for a seed-eater. Since, however, there are more than a dozen small islands in the Galapagos group, these finches had an opportunity to develop in relative geographical isolation. Variations which may have come about by the conditions described above do not necessarily make new species, one of the criteria being that different species can no longer interbreed. What probably happened on the Galapagos Islands was that a number of sub-species evolved in isolation on a number of islands. If one sub-species spread to an island occupied by another race of the same species, and the two populations had not been isolated for long enough and differed only in minor ways, they would then probably interbreed freely and merge with each other.

Evidence has now accumulated suggesting that if two populations of the same species have been isolated for a long time, enough hereditary differences may have accumulated to prevent the genes from combining well. Such hybrid offspring would not have a good chance of survival and natural selection would tend to intensify the gap between the two parent forms allowing them to develop into two separate species when eventually interbreeding becomes impossible.

FIG. 49. Twelve of Darwin's finches of the Galapagos Islands. This group of birds became adapted to diverse modes of life and developed a great variety of individual features, particularly in the structure of the beak. (After David Lack.)

So then, after a long period of time during which the Galapagos finches evolved in their various island homes, it eventually came about that different sub-races spread from one island to the next only to evolve further in the particular environment that suited them. The result is that today as many as six different species which no longer interbreed can be found on one island. In the words of Professor David Lack: 'The differences in the beaks that can be found among the Galapagos finches are not just an insular curiosity, but strict adaptations to differences in diet and are an essential factor in their ability to live together.'

This, then, is an example of how by a process of natural selection and geographical isolation new species formed from what originally might well have been only one species from the South American mainland.

The Ring of Races

IN THE MOUNTAINS of California there live two races of salamander which do not interbreed. The American herpetologist, R. C. Stebbins, studied these creatures, one of which was light and even in colour, while the other was boldly marked. He traced the light variety to a coastal area in Southern California, and a little farther to the north he found a darker race. Farther north still, in Northern California, a much darker and evenly coloured race extends its range far beyond the coastal regions into the interior, only to be replaced to the south-east by a mottled form. Farther south in the hot and dry climate of the Sierra Nevada this mottled form is yet again replaced by a different mottled variety. The ring of salamanders is completed by another boldly marked salamander which overlaps the range of the evenly coloured form in Southern California.

The extraordinary thing about this ring of salamander races is that neighbouring sub-species can interbreed with each other; that is A can interbreed with B, B with C, C with D, D with E, and E with F. But, where the ring closes and the breeding areas of F overlap with A, no more, or only very rare, interbreeding takes place. If, for some reason such as geographical changes in the land surface, the ring of connecting races were to be broken, the now-continuous species of salamander with its distinct, non-interbreeding ends would have to be regarded as two separate species instead of one.

A number of other rings of races, particularly among birds, are known, and there are examples where, during the course of evolution, a ring of races has become a ring of similar species which no longer interbreed with each other.

Why Different Species Do Not Interbreed

EVEN if two animal species are of similar appearance they can be classified as different species providing they do not interbreed. There are good reasons why animals of different species cannot successfully interbreed. For example, it is possible to fertilize an egg of a bullfrog with a sperm of the leopard frog. Although the resulting zygote starts to divide, the embryo soon dies. Another result of inter-species crossing can be a vigorous but sterile hybrid. The mule is a well-known example. The crossing of a horse and a donkey results in a mule.

Very common, particularly among birds, are examples of different species which are able to produce vigorous and fertile hybrids if they are crossed artificially, but which will not interbreed in nature. The reason for this is that different bird species often have very different nesting habits and mating instincts. Thus, the male birds of the grouse and woodcock perform elaborate courtship displays in front of the females before the latter are ready for mating. This display seems to stimulate the secretion of sex hormones in the hen bird, resulting in the release of her eggs. The hen birds are only strongly stimulated by the display of males from their own species, and thus little or no mating with males from other species takes place.

In nature, the formation of new species is normally a very slow process which may take thousands of years to complete. But there is a way by which new species can be formed immediately after the crossing of two existing species. This happens when the number of chromosomes in the hybrid zygote is doubled when the sex cells (gametes) fuse.

In 1926 the Russian scientist Karpechenko carried out such experiments when he crossed a radish with a cabbage plant and obtained a vigorous hybrid. When, however, this plant produced flowers he found them to be completely sterile. The reason for this hybrid's sterility was that the 9 chromosomes derived from the radish simply did not match the 9 chromosomes derived from the cabbage. In further experiments Karpechenko discovered that sometimes the sex cells of the hybrid (the pollen grains and egg cells) retained the full number of 18 chromosomes, 9 from the cabbage and 9 from the radish. In fact, the usual halving (meiosis) of the chromosome numbers did not take place in the formation of the gametes. If these gametes fused they gave rise to a vigorous hybrid plant with 36 chromosomes, double of either the cabbage or the radish. Offspring from these hybrids were fertile and bred true if crossed among themselves. Karpechenko had thus created the first artificial species. Since those days similar experiments have been carried out successfully with other plants, good examples among garden flowers being the cultivated marigold and larkspur.

The Record of Life

WE HAVE in the last chapter surveyed the possible beginnings of life.

From a practical point of view, however, it is mostly the last 600 million years of the earth's history which are of interest to the palaeontologist, because before this, during Pre-Cambrian times, traces of life are exceedingly rare. It is however noteworthy that in recent years the Americans P. E. Cloud and J. Schopfer as well as others have devoted a great deal of attention to Pre-Cambrian fossils and much progress has been made in their elucidation. Nature's chemical experiments which occurred on our planet at the beginning of life are still largely speculative, whereas the evidence for the experimental processes of evolution drawn from palaeontology rests on detailed and abundant fossil records capable of explaining much of what happened to species during the course of millions of years.

As fossil records show, the history of life seems closely related to the physical changes which took place on the surface of the earth. As continents and seas changed and climates altered, life was repeatedly modified in response to the new environments, but interrelationships between the the two are often complex and difficult to decipher.

In most cases only the hard parts of animals are capable of preservation as fossils. Soft parts are fossilized only under exceptionally favourable conditions as in frozen soil, ice, amber and petroleum-saturated soil. Sometimes the soft parts are represented by an impression upon the surrounding rock, by stains and even by petrification likely to preserve the microscopic structure of muscles and other tissues. The preservation of any animal or plant fossil depends on its burial in a sediment that is not subsequently destroyed by denudation.

In the sea the remains of marine animals are continually accumulating on the ocean bed. They may lie where they fall or be carried considerable distances by sea currents until eventually they come to rest, often in association with other objects of similar weight. Thus it is common to find considerable accumulations of fossil shells of approximately the same size which are all some distance from the place where the organism originally lived. Some species, such as the shells of the burrowing molluscs, may be found where they were in life, but the remains of animals in the sea can only be preserved if they are deposited at a place where mud and sand is accumulating. If they remain exposed for long periods the action of the sea destroys them.

Land fossils are even more vulnerable and their survival is usually due to a succession of lucky accidents. The remains of land animals are chiefly preserved either in the sediments that form on the bottom of lakes and river beds or in arid areas where wind action raises the general level of the land surface by depositing dust and sand. Under the latter conditions the skeleton of a dead animal is likely to become covered by blown sand and may remain buried for several geologic periods. Animals that die in a river bed or on a lake shore often quickly become buried in sediment, the water preventing attacks by other organisms. Under special circumstances the material of an animal's skeleton may be removed and replaced by mineral matter. This process is commonly accomplished molecule by molecule so that even microscopic structures are preserved while the object is literally turned into stone.

The ash of erupting volcanoes frequently buries all surrounding animal and plant life. These organisms too may be preserved as fossils. In Poland entire carcasses of the extinct woolly rhinoceros are preserved in oil seeps, and the famous fossils of Rancho La Brea in the city of Los Angeles have been preserved in tar pits.

Extinct mammals of the Ice Age such as the woolly mammoth have been found in the ice or frozen ground of the Arctic where they remained under deep-freeze conditions for thousands of years. The best known of these is the Berezovka mammoth found in Eastern Siberia in 1899. It was virtually complete, with clotted blood in the chest and unswallowed food in the mouth.

Some fossils are remarkable for the exquisite detail they reveal. Such examples stem from the Middle Cambrian period, 550 million years old, of the Burgess Pass shale, British Columbia. Here the external form of slender soft processes of worms has been perfectly preserved.

A well-known example of a somewhat different type of protection is that of insects, flowers and leaves preserved in amber, which is itself the fossil resin of extinct pine trees. This has long been known from the

Tertiary rocks of the Baltic. The insects became trapped in pine resin which later hardened around them. They are preserved as a carbonaceous film which faithfully reproduces even the minutest details of their external structure. Usually these fossils consist only of perfect external moulds, although occasionally the muscles have also been preserved.

A natural occurrence especially important in the preservation of plants is the detritus of forests growing on a coastal plain at or below high water mark. If such a plain is situated in a sinking area, great masses of deposits (which form many of today's coalfields) can be built up. In these, animals and particularly plants will be preserved.

The Dating of Fossils

Fossils are studied to elucidate trends in evolution, and to discover possible relationships between fossil animals, e.g. whether certain animals have been ancestors to others. Fossils also date the rocks in terms of geologic time. Once the sequence of fauna and flora has been worked out in a long succession of undisturbed sedimentary deposits, it becomes possible to determine the relative ages of the fossils by the way they lie superimposed, the oldest formations being the lowest and each in turn being younger than the one on which it rests. No single region of the world contains a complete record of all geologic time, and even if it existed, the lower part would probably be so deeply buried as to be inaccessible. It has therefore been necessary to assemble the evidence from many different parts of the earth, by correlating partially overlapping sequences. In this way geologists throughout the world have built up the so-called Geologic Column, and a corresponding Geologic Time Scale.

Fossils also frequently indicate the local environment under which many of the stratified deposits have been laid down. The presence of corals and other marine organisms proves, for example, that the enclosing sediments were laid down on the sea floor, although such fossils are frequently found far inland and even on high mountains. On the other hand, the presence of fossilized skeletons of land animals and of tree stumps in their place of growth can be taken as evidence that the area was land when deposition took place.

A number of methods have been developed for dating fossils. These are based on the measurement of amounts of elements which, in the process of time, are either formed by or are subject to radioactive decay.

Radiocarbon dating is one of the most important and accurate of these methods. All forms of life contain non-radioactive carbon C^{12} and a very small quantity of the radioactive kind C^{14} continually produced in the upper atmosphere by the interaction between atmospheric nitrogen and cosmic rays. C^{14} quickly combines with atmospheric oxygen to form radioactive carbon dioxide, $C^{14}O_2$. This, together with the non-radioactive form $C^{12}O_2$, is taken up by plants and animals.

Because the quantity formed is balanced by the quantity which decays, the amount of C^{14} present in the world remains more or less constant. This also means that a fixed, very tiny proportion of the total carbon present in living organisms is of the radioactive kind. When an organism dies, however, no fresh radioactive carbon can be taken into the system, while that already present will decay at a constant rate. Measurements have shown that, during each interval of 5,570 years, the radioactivity of C^{14} is reduced by exactly half. This time span is known as the half-life of C^{14}. Thus, by measuring the amount of C^{14} present in animal or plant remains their age can be determined with accuracy up to about 50,000 years. If, for example, an animal fossil is found to contain a quarter of the normal quantity of C^{14} it means that this animal died about 11,140 years ago. The amount of C^{14} remaining in fossils older than 50,000 years is so small that its measurement is too difficult and inaccurate to be of use.

The potassium argon dating method provides another means of dating fossils. It is based on the fact that the radioactive isotope potassium^{40} decays at a constant rate to calcium^{40} and the gas argon^{40}. By measuring the ratio of argon^{40} produced to the remaining potassium^{40}, the age of potassium-bearing rocks can be so determined. When this method is used for dating fossils, these should be closely associated with the rocks under test or be situated in the same geologic strata. It must be assumed of course that most of the argon^{40} gas was trapped in the rocks, and certain corrections are also necessary to allow for contamination by argon^{40} from the air. Because of these factors the method is unsuitable for rocks younger than about 400,000 years. It can, however, be used successfully on rocks dating back to and beyond Cambrian times (over 600 million years before the present).

Another important method for determining the age of rocks is that based on the radioactive decay of thorium and uranium to lead. As with the potassium argon method, fossils to be dated should be closely associated with the rock under test. The minerals are carefully analysed and the proportion of the decay products to the original substances is determined.

A recently developed chemical method useful for the relative dating of animal bones and teeth is known as the fluorine test. Bone contains calcium phosphate which combines with fluorine found in ground water. Thus, by measuring the fluorine to phosphate ratio in skeletal remains found in the same deposit, it is possible to ascertain whether the fossils being investigated are all of the same age. If so, they should have accumulated similar relative amounts of fluorine from the ground water, and their fluorine–phosphate ratios should be constant.

This method is also useful for determining whether fossil bones and teeth found together belong to the same animal. When using the fluorine test, however, account must be taken of the fact that bone and antler absorb fluorine more readily than teeth. Thus, even if of the same age, the former will contain a higher proportion of fluorine than the latter. These and allied methods have been used to establish a number of evolutionary sequences, amongst which those of the horse, camel, giraffe and elephant are of particular interest.

The existing fossil record of the history of life on earth reveals that life has been gradually changing from one form to another over millions of years. It now provides overwhelming evidence that organic evolution has occurred through past ages and that it remains the principal mechanism by which organisms adapt and change.

The Geological Calendar

PALAEONTOLOGISTS have devised a geological calendar which attempts to set out the relationship between physical changes on earth and those in living forms. The largest divisions in this calendar, the eras, correspond reasonably well to time intervals between intense upheavals in the earth's crust. The names of eras end in the suffix 'zoic', which stems from the Greek *zoos*, meaning 'living'. Thus Palaeozoic means ancient life, Mesozoic middle life, and Cenozoic refers to recent life. Every era has its characteristic pattern of life. Geologists have also named the major layers of rock in the earth's crust, and their ages have been estimated by various radiometric means. These, then, represent the periods and lesser units in the geological table, each having distinctive assemblages of fossil animals and plants. Scientists soon found, however, that as they started to study

THE SUCCESSION OF LIFE THROUGH GEOLOGICAL TIME

Era	*Periods*	*Epochs for Cenozoic Era*	*Plants*	*Animals*	*Climates*
	QUATERNARY	PLEISTOCENE 3,000,000 years ago	Increase of herbs Decrease of trees	Human cultures Extinction of many large mammals Early man	Distinct zones of climate and changes of season
		PLIOCENE 13,000,000 years ago	Rise of herbs Increase of grass-lands	Mammals now widespread Pre-human hominids	Cool and temperate climate away from the equator
		MIOCENE 25,000,000 years ago	Spread and develop-ment of grasses Decrease in forests	Mammals on the increase	Climates get cooler
CENOZOIC	TERTIARY	OLIGOCENE 36,000,000 years ago	Tropical forests widespread over the whole world	The appearance of modern mammals Early anthropoids	Warm climates probably still wide-spread over the world
		EOCENE 58,000,000 years ago	Extension of plants possessing seeds enclosed in capsules (angiosperms)	Older groups of mammals Early primates	Well-established belts of climatic zones
		PALAEOCENE 65,000,000 years ago	Modernization of plants possessing seeds enclosed in capsules (angiosperms)	Rapid develop-ment of mammals	Development of climatic zones
	CRETACEOUS	135,000,000 years ago	Development of flowering plants Decrease of conifers	Decrease of reptiles	Climates begin to diversify
MESOZOIC	JURASSIC	180,000,000 years ago	Conifers and cycads widespread Beginning of the flowering plants	First birds Spread of reptiles Abundance of dinosaurs	Climates are warm all over the world

	Triassic	230,000,000 years ago	Increase of conifers First known cycads	Transition of reptiles to mammals Rise of reptiles	Worldwide tropical and sub-tropical climates
	Permian	270,000,000 years ago	Rapid decrease of ancient plants Seed ferns extinct	Reptiles become dominant Amphibians still very important	Varied climates
	Carboniferous	350,000,000 years ago	Great tropical forests First mosses, seed ferns and conifers	Amphibians dominant on land Rise of insects	Uniform climates Spread of tropical seas
	Devonian	400,000,000 years ago	First forests First known liverworts, horse-tails and ferns	First amphibians Wide radiations of fishes	Uniform climates Rise in temperatures
Palaeozoic	Silurian	440,000,000 years ago	Appearance of first land plants Club mosses	Invertebrates highly developed	Climates slightly cooler
	Ordovician	500,000,000 years ago	Marine algae dominant	First appearance of fishes Invertebrates highly developed	Climates warm
	Cambrian	600,000,000 years ago	Algae, fungi, bacteria	Invertebrates widespread and varied	Climates becoming warmer
	Pre-Cambrian	4,500,000,000 years ago (probable age of the earth)	Algae, fungi, bacteria	First primitive invertebrates	Climates cool at the end of Pre-Cambrian

The table gives the approximate beginnings of Epochs and Periods.

smaller and smaller units, local effects such as climate and geographic position added importance and made precise dating more difficult.

The geological calendar, built up in a rather haphazard manner over a period of some 250 years, had been well formulated before any possibility of determining the absolute ages of rocks in years existed. Despite its shortcomings, however, it is an invaluable guide not only to an overall understanding of the evolutionary process, but as a framework for evolutionary studies.

4
The Evolution of Man

IF WE EXCLUDE the micro-organisms—protozoa, bacteria and viruses—more than a million known kinds of animals live on the earth. Roughly half of these are insects, and of the remaining half-million animals, about a fifth are vertebrates (animals with backbones). Of those, only some twelve thousand are mammals.

In our present age it is apparent that mammals are the dominant animals on earth—partly because man has outdistanced all other creatures by the extraordinary development of his brain.

One might legitimately wonder whether under today's earth conditions other animals—particularly primates—would have been as successful as man, had he never existed. The answer is probably no. A good deal of evidence is available to show that our nearest relatives, the apes, are on a path to extinction, and not only because of man's control over their environment and his general tendency to destroy wild life. Indeed, Dr George Schaller's extensive studies of gorillas in the wild suggest that these fine animals may, to some extent, lack reproductive drive. In 466 hours of observation he witnessed only two matings. Reasons for the evolutionary success of species include successful reproduction, adaptability to changing environments and, not least, a genetic constitution which will allow animals to pass through their years of fertility without succumbing prematurely to disease.

As far as variety and size are concerned, the zoological class Mammalia is hard to beat. It includes almost all the larger animals inhabiting the surface of the earth as well as the whales and seals of the seas and the bats in the air.

What are some of the main characteristics that have assisted the evolutionary success of the mammals?

Successful reproduction is a prerequisite for any group of animals destined to become widespread and dominant. Through the process of natural selection, the mammals solved this problem largely by discarding the wasteful egg-laying process of the reptiles (there are only two Australian types which still lay eggs), and adopting a new method by which they brought forth their young alive and suckled and protected them during their helpless infancy.

Mammals, too, are warm-blooded animals, able to maintain an even, often fairly high, body temperature. There are also some exceptional mammals which hibernate, when special mechanisms allow for a lowering of body temperature during the dormant phase.

Another evolutionary adaptation of mammals, to preserve their body heat, is the development of a hairy covering to protect them from temperature extremes.

Although even the most primitive mammals show advances in brain structure over the reptiles, their most important advance in brain development has been an increase in size and complexity of the surface areas of the hemispheres known as the cerebral cortex ('grey matter'). The intelligence of mammals does not appear to be related so much to the whole bulk of the brain, but rather to this cerebral cortex where, in the more progressive animals, convolutions add vastly to the total surface area. Particularly is this so in the primates, which include man. Brain size also varies to a degree with body size, bigger animals having larger brains than smaller ones without necessarily being more intelligent.

The Reptilian Ancestors of Mammals

TRACES of mammal-like skeletons of reptiles date back to the late Carboniferous and Lower Permian periods (270 million years ago). These so-called pelycosaurs, some of which were ferocious meat-eaters, left skeletal remains in the Texas redbeds. The first approach to mammal structure is noticeable in their skulls. Here, instead of a uniform tooth row common to the reptiles, there are a few front teeth, comparable to incisors, usually two large canines, and at the back of these, a series of teeth corresponding to the molars and premolars of mammals.

Fossils of mammal-like reptiles have been found in the Karroo beds of South Africa, descended from the pelycosaurs, which date back to the Permian and Triassic periods (230 million years ago). Among these so-called therapsids were a number of aggressive meat-eaters with skull forms intermediate in type between those of the primitive reptiles and mammals. Their teeth and limbs were both approaching mammalian conditions. The former showed a differentiation into incisors, two single canines, and molar-like teeth, which were sometimes cusped instead of being primitive reptilian cones. The front and hind limbs had shifted markedly to the front and back, the mammalian position. The bones, too, showed modifications towards mammalian skeletons. The therapsid skeletons, however, fail to tell us whether these animals were already warm-blooded and nursed their young. Neither do we know if they had hair or scales. Although they are grouped as reptiles, if we could see them alive they would probably remind us more of a mammal than any of the living reptiles of today.

These animals, the commonest of reptiles during the later Permian and the early Triassic periods, did not, as one might expect, give rise directly to the large mammals of today. On the contrary, most of them disappeared during the end of the Triassic, being crowded out by the archosaurs, the ruling reptiles, which included the dinosaurs and flying reptiles—animals that dominated the world stage for a hundred million years or more during the Mesozoic era.

Unfortunately, fossil traces of the first primitive mammals are scarce and our knowledge of them depends almost entirely upon teeth and jaws. It seems likely now that the transition from reptiles to mammals took place at the beginning of the Mesozoic era, about 180 million years ago, during the Triassic period. This, as mentioned, was also the time when the ruling reptiles, the dinosaurs and others, were dominant animals on earth. Not all dinosaurs were giants, many of them being quite small, but there were fierce meat-eaters among them and this made life difficult and dangerous for the first mammals which, on fossil evidence, appear to have been no bigger than rats or mice. By necessity, they must have been of retiring habits, making their nests in small burrows, whence they would emerge under the cover of darkness to feed on a varied insect and vegetable diet. Their fossil skulls reveal sharp teeth, indicating flesh-eating tendencies, but probably they were too small to kill other vertebrates and thus had to content themselves with insects and worms supplemented by a variety of vegetarian foods. Their brains, although still very primitive and poorly developed, already showed certain advances over those of reptiles.

During the Triassic and Jurassic periods the world climate is believed to have been generally very warm. Warm-bloodedness, therefore, was not so important to mammals as it became in later geological periods when parts of the world became extremely cold, excluding many reptiles from such habitats and bestowing great advantages on those animals able to regulate the temperature of their bodies internally.

Even in those early days, however, when world climates were generally warm, warm-bloodedness must have allowed the first mammals to invade areas at high altitudes where low temperatures did not permit cold-blooded reptiles to survive.

In the late Mesozoic, some 65 million years ago, conditions on earth changed to become unfavourable to the huge reptilian population still prevalent at the time. The evidence of fossils shows that these creatures, which for over a hundred million years had dominated the fauna in vast regions of the earth, began to die out in enormous numbers. Within a few million years only a few remained in some of the warmer climates, and it is from these that today's known species of reptiles have evolved. The wholesale disappearance of the reptiles must have considerably increased the food supply available in various habitats for the yet small and primitive mammals. Whatever the reason, however, mammals increased in size, variety and numbers during the Palaeocene, soon after the bulk of the reptiles had disappeared.

Just as the world had become dominated by reptiles in the Triassic and Jurassic periods, the Palaeocene saw the mammalian population rapidly increase and branch out into many different groups. Some of these were of large size, like elephants, whales, lions and rhinos. A few again became specialized as burrowers or gnawers, similar to the modern rodents. Others still had evolved into fierce little meat-eaters like the stoats and weasels of today. All this meant that living space in many habitats became overcrowded, and competition between the various species for food and suitable breeding territories must have been fierce. Little wonder, therefore, that some of the smaller ground-living mammals took to the trees, where they could find a plentiful supply of insect food or, for those that preferred it, a vegetarian diet of leaves and fruit.

It was at this stage of evolution that the first experiments with the primate body began.

The Primates

TO THE PALAEONTOLOGISTS primates are a source of considerable difficulty because there is still a lack of knowledge as to the many details of the structure of the earliest primates. Well-documented pedigrees have been constructed for many of the common mammals such as horses, camels and dogs. Unfortunately, as far as primates are concerned, their fossil record remains inadequate perhaps because primates have always been for the most part forest dwellers, and rock deposits in which fossil vertebrates are to be found are not normally formed in forested regions.

Considerable imagination would be needed to accept the idea that man has evolved from a small creature with a shrew-like snout fitted with whiskers, four upper and six lower front teeth, five sharp-clawed fingers and toes, and a long bushy tail of which the fur on the under side is short and harsh. This description fits the tree shrews of today, of which there are five living genera widely distributed in South-East Asia. They often hold food in their hands and sit up to eat. Their diet is completely omnivorous and some of their structural features and much of their general behaviour points towards the great primate evolution.

FIG. 50. Living tree shrew.

This, of course, does not mean that we have evolved from any of the existing species of tree shrews, but long before the giant dinosaurs became extinct the ancestors of the mammals were showing variations from the other reptiles as the result of natural selection. Very early in the evolution of mammals a line of insect-eating shrews, which we know today only from fossil records, such as *Plesiadapis*, a tree shrew of the Palaeocene, took to the trees. From these small mammals, many of whose features are very similar to the tree shrews living today, the first primates evolved.

One of the striking features of the primate order, wherein it differs from other mammalian orders, is that its existing members fall into a graded scale of organization. Quite apart from the evidence of the fossil records, this suggests an actual evolutionary change within the order, leading from the most primitive, the tree shrew, to the most advanced, man. T. H. Huxley wrote:

> Perhaps no order of mammals presents us with so extraordinary a series of gradations as this—leading us insensibly from the crown and summit of the animal creations down to creatures from which there is but a step as it seems to the lowest, smallest and least intelligent of the placental mammals.

Man has become the dominant mammal because of his brain, and as we shall see later his brain was the consequence of being a primate and a descendant of primates. There is, of course, a vast gulf between a primitive tree shrew and *Homo sapiens*, but in evolution's experiments time has often been on the side of the experimenters. And there was plenty of time. It took us around 70 million years to evolve from the insignificant state of a small tree-climbing insectivorous mammal to *Homo sapiens*, the primate able to launch himself into space.

An important feature, universal among primates, which was vital in allowing us to emerge as sophisticated mammals, is the ability to grasp with our hands. In contrast with such arboreal mammals as the squirrel, which climbs by digging its claws into the bark, the primate hold is generally accomplished by grasping branches or twigs. In most of them, the primitive claws have evolved into flat nails serving as a protection to the finger-tips, but there are exceptions. Some lemurs and the marmosets have either retained claws or returned to a claw-like structure.

Man's hand, like that of other primates, has retained the five digits (fingers) of the earliest four-footed animals, the three segments of each finger, with the exception of the thumb, being typical in mammals. The thumb is capable of being rotated inward from the palm, allowing it to be set against the other fingers and wrapped around an object from the other side. Human thumbs can thus be used efficiently in opposition to the other fingers, a vital characteristic which makes our hands capable of sophisticated manipulations. The adaptation is particularly advanced in man. Other primates also have opposable thumbs but in their case it is more apt to be the great toe which has better opposition to the other toes.

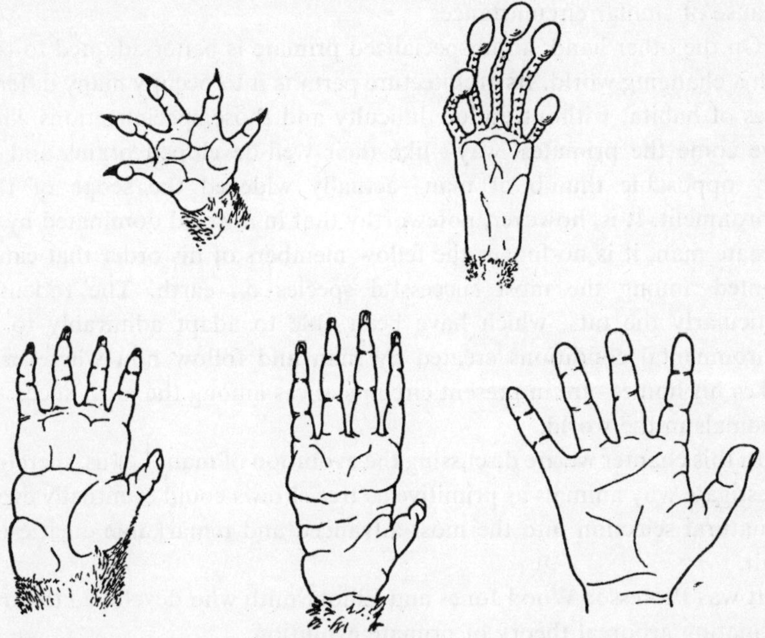

FIG. 51. Some examples of primate hands. *Top* Opossum. Tarsier. *Bottom* Baboon. Orang-utan. Man. (After illustrations in the British Museum (Natural History), by permission of the Trustees.)

Except for their brains, which in higher primates especially have become highly organized, the primate body has retained many unspecialized features quite unlike other orders of mammals. This evolutionary trend may well have been the primate's good fortune. For while the hoofs of horses enable these animals to attain high speeds on flat terrain, they also limit the kind of environment to which they can adapt. Throughout the later part of their evolutionary history the habitat of horses was open grassland. If, for example, a changing climate such as severe drought or cold should make grasslands uninhabitable for horses, by reducing the food supply, the animals might be driven into rocky mountain areas where their specialized feet would make life difficult for them even if there were sufficient food available. Thus sudden climatic changes may leave insufficient time for adequate evolutionary adaptation to take place, and highly specialized species of animals may face extinction. This is but one example of what might happen to an over-specialized organism if its environment altered severely. Probably many animals became extinct because of similar circumstances.

On the other hand, an unspecialized primate is better adapted to cope with a changing world. Its architecture permits it to occupy many different types of habitat without undue difficulty and those specializations which have come the primates' way—like their well-developed brains and the fully opposable thumb of man—actually widened the scope of their environment. It is, however, noteworthy that in a world dominated by the primate man, it is no longer the fellow members of his order that can be counted among the most successful species on earth. The rodents—particularly the rats, which have been able to adapt admirably to the environmental conditions created by man and follow him wherever he makes his home—are in present circumstances among the most successful mammals in the world.

In this chapter we are discussing the evolution of man. Let us, therefore, investigate why animals as primitive as tree shrews could eventually evolve by natural selection into the most advanced and remarkable creature on earth.

It was Professor Wood Jones and Elliot Smith who developed the truly fascinating arboreal theory of primate evolution.

To a small mammal like the tree shrew the lofty tree tops are its whole world. Not a flat world to which we and many other mammals have become accustomed, but three-dimensional, with vertical and horizontal pathways. According to the arboreal theory, one of the earliest stages in primate evolution was a slightly rotating thumb and big toe able to oppose the other fingers and toes. Thus, even in the most primitive of primates, the tree shrews, the first digits on paws and feet can already be widely separated from the other four. A very different state of affairs is found in terrestrial animals where the natural working axis of the feet runs along the middle, decreasing the importance of the digits to either side so that they have become much reduced in animals like cattle and pigs and were even totally lost during the evolution of the horses.

Thus it came about that tree-living animals used their fore limbs more and more to explore, while the hind limbs became the principal supporters of the body. Professor Wood Jones has termed these changes, 'the differentiation of the limbs'—the more the hind limbs were able to accept the duty of supporting the body, the more the fore limbs could be emancipated for other diverse uses. Thus, eventually, the early primates were able to sit upright in their environment of trees and branches, perching on their hind legs and using their arms and hands to investigate and hold objects.

In ground-living quadrupeds the snout serves a variety of very important functions. It accomplishes, in fact, almost everything concerned with eating. The fore limbs, apart from transporting the animal to its feeding place, have little further function. The food is broken up and chewed almost entirely by means of the jaws and teeth. It was, therefore, of considerable evolutionary value for ground-living mammals to develop an extended snout. It enabled them to use their jaws for effective fighting, but more important, it provides mammals like the wolves with the only means of killing and crushing their prey.

Some authorities have suggested that a snout is a useful adaptation to allow an animal to use its eyes to see what it is doing with its own snout. Whether mammals in general are so short sighted that they can focus adequately at such close distance seems doubtful. Man, with his excellent sense of vision, certainly finds it difficult to focus comfortably closer than seven or eight inches.

The environment of the tree tops, of course, makes a different demand upon the senses of an animal. The ability to smell loses a great deal of its value because, unlike unbroken ground surfaces, the branches of a tree provide less area to which scent can cling. Moreover an animal making its way through the branches of a tree cannot comfortably sniff at everything that arouses its interest without running the risk of accidentally falling from the tree. Thus, as an instrument of touch, the muzzle has lost much of its importance among arboreal creatures. It is far easier for them to investigate their surrounding world by using their hands. Foliage, fruits and insects can be readily grasped and investigated at convenient distances from the eyes instead of only at the fixed focus available to, say, a creature like the wolf which, when it feeds, has no means of changing the distance between eyes and snout. Thus, the use of hands together with stereoscopic vision makes for a far better co-ordination between sight and touch.

The environment of trees was also directly responsible for the evolutionary diversion which led primates from smelling with snouts to feeling with their fore paws. The functions of the jaws gradually changed simply to chewing, leaving the activities of catching and killing the food to the fore paws. There are some living monkeys, notably the baboons, that have a very much enlarged muzzle, unlike that of most other primates. This seems to have been a subsequent evolutionary development which also provided these animals with enormously powerful jaws and huge canine teeth—a very necessary defence mechanism for these ground-living species. It was, therefore, life in the trees which opened up new vistas for the developing hands of primates. Grasping paws and sensitive fingers played a vital role in the evolutionary development which eventually led to man.

Dr G. G. Simpson's scheme of classification of the order of primates is today the generally accepted one. According to this classification, the primates are divided into two sub-orders, the *Prosimii* (or lower primates) and the *Anthropoidea*. The latter includes man, anthropoid apes and monkeys, while the former includes tarsiers, lemurs and tree shrews.

The Prosimians

THE FIRST definite primate types are known to have existed in the Palaeocene, some 70 million years ago. These and later prosimians were primitive in the sense that they retained many of the characteristics of insect-eaters, viz. rather long faces, eyes placed laterally and somewhat small brains. The precise time of the origin of the *Anthropoidea* is still very much in doubt, but fossils of these primates are definitely known from the Oligocene some 40 million years ago.

Before we delve further into the past to investigate possible connections between ourselves and the early primates, let us familiarize ourselves with members of our order still living today.

We have already met the tiny tree shrews which occupy so low a rung in the ladder of our order that some zoologists have rejected them as primates and classed them with the insectivores. Yet that part of their brains serving sight (the cortex) shows a slight increase in size while that connected with the sense of smell is reduced. The skull, too, shows certain primate features and consequently most authorities today allow them to remain in our order.

The next step in primate evolution brings us to the lemurs, nocturnal creatures most of which are much larger than the tree shrews. Still well down in the primate scale, they have a fairly long muzzle but their eyes, pushed farther to the front of the face, are considerably larger than those of the tree shrews. They spend their time climbing among the branches of trees, helped by typical primate five-fingered hands and feet. Their fingers are fitted with nails, but the second toe bears a claw. Lemurs have unspecialized teeth, except for the incisors and the canines of the lower jaw which form a characteristic comb. Their dentition is well suited to a varied diet. They are a very ancient group of mammals and fossils of their ancestors, some of which date back to Eocene times, have been recovered from America and Europe. About 40 million years ago, during the Oligocene, their fossil records disappeared from Europe and North America and evidence shows that Asia and Africa—where a number of species still survive today—became their new homes.

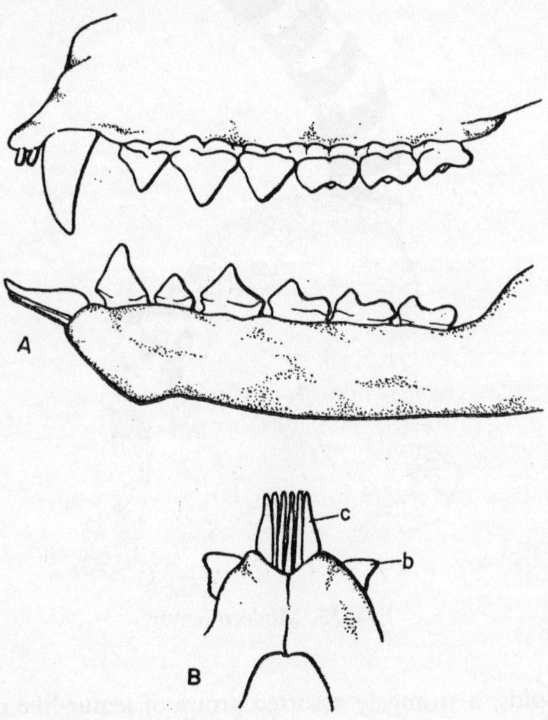

FIG. 52. The dentition of a lemur, side view (A). The upper incisor teeth are much reduced. The canine tooth is sharp and dagger-like. Part of the lower jaw (B) showing the 'dental comb' where the canine (c) has been curiously modified. The front premolar (b) assumes the function of a canine. (After illustrations in the British Museum (Natural History), by permission of the Trustees.)

So-called true lemurs live on the island of Madagascar, where they have evolved for the last 50 million years. About forty species are found on the island, including the spectacular black-and-white Indri, which stands thirty-nine inches high, and the mouse lemur *Microcebus*, which is only the size of a fist.

FIG. 53. Modern lemur.

The lorisoids, a strangely assorted group of lemur-like mammals, are creatures that, on a great deal of demonstrable evidence and especially that of fossil types, have long been separated from the true lemurs. The group includes animals of very different temperaments and appearance—the slow-moving lorises contrasting sharply with the agile bush-babies. The largest of the lorises, the slow loris, measures up to sixteen inches in length and is almost tail-less. It is still found, in ten distinguishable forms, in Eastern Asia and in the islands of Sumatra, Java and Borneo. Its thumbs and great toes are widely opposed and point almost directly away from the four fingers and other toes. The index finger is small and the thumb very much enlarged. All the digits bear small nails except that adjacent to the big toe which, as in all *Lorisidae*, is fitted with a large recurved claw. Nocturnal in habit, it is omnivorous and feeds on a wide variety of animal and vegetable foods. The slow loris has been observed catching insects with its hands while suspended from a branch by its feet. This is true primate behaviour—where an animal uses its fore limbs for operations other than the simple support of the body.

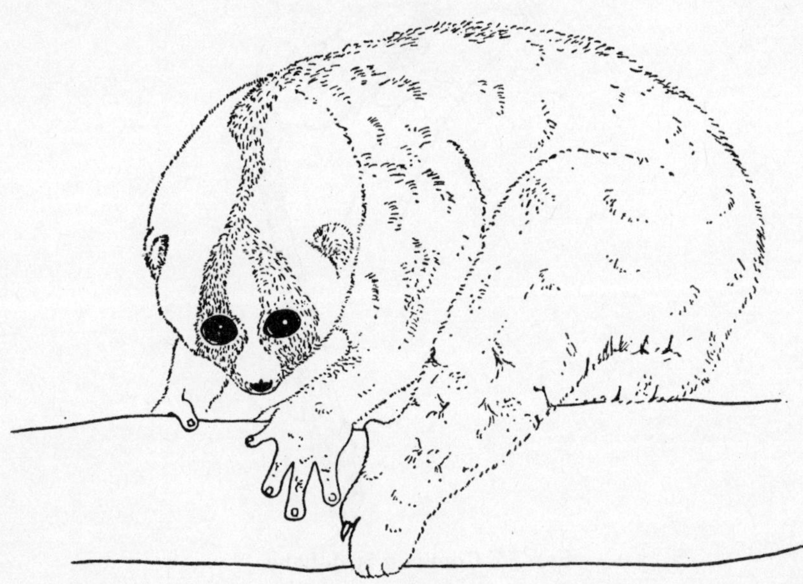

FIG. 54. Slow loris.

Bush-babies, of which three distinct kinds are known, possess long bushy tails and very large ears. Four varieties live in Africa and the principal type, the common bush-baby, occurs widely throughout the forested regions of that continent. Bush-babies are also characterized by their large, staring eyes, the pupils of which can contract to almost invisible pin-points. They have large thumbs and big toes. They are also opposed, but not to the extent of the lorises. The fingers and other toes are long and slender, provided with large, fleshy terminal clinging pads.

Most bush-babies are nocturnal in habit and eat largely vegetable matter. Agile little creatures, they spend their entire waking lives flitting about in the canopy of trees.

FIG. 55. Common bush-baby.

An odd and somewhat grotesque looking prosimian is the tarsier, sometimes described as a living fossil. It dwells in the great equatorial forests of the Indonesian islands. Active and skilful, it spends much of its time perched upright in bamboo groves. Its grotesque appearance is the result of two enormous eyes set into a round, short face. It also has unusually mobile and rather bat-like ears. Specially adapted joints between the skull and the neck vertebrae allow the head to be turned through one hundred and eighty degrees so that it can face directly backwards.

The tarsier is a nocturnal insect-eater. It catches its food with its hands and brings it up to the mouth to eat. Its limbs are specialized for grasping and hopping and the fingers and toes are all very long and thin, ending in bulbous fleshy pads shaped like the discs on the feet of tree frogs. Tarsiers are the sole survivors of an animal group now long extinct. In Eocene times, however, between 60 and 40 million years ago, many of its early relatives inhabited forests over wide areas of the world, including Europe and America. Some of these extinct forms were very like the modern genus, *Tarsius*, but others seem a good deal more primitive, with smaller brains and less specialized development of the limbs.

FIG. 56. Tarsier (*Tarsius spectrum*).

The evolutionary origin of the tarsiers is still in doubt. Some zoologists believe they may have been derived from very primitive and generalized forms of lemurs, before the ancestral lemurs developed those specialized structures which distinguish modern lemurs from other primates. Others think it more probable that tarsiers descended directly from tree shrew-like ancestors without passing through a lemur phase. There is, however, general agreement that in Eocene times there came into existence numerous kinds of small tree-living mammals, not unlike the tree shrews of today, from which were derived several types of primitive lemurs and tarsiers. These active primates were a successful group of animals and soon spread over a great part of the New and the Old World, many of them representing side lines of primate evolution which eventually became extinct. Fossil evidence further reveals that some of the tarsiers showed a progressive trend in their development which provided the basis for further evolutionary advances, later leading to the appearance of true monkeys.

Although it takes only a few words to describe these evolutionary changes, it is as well to remember that the Eocene period lasted about 25 million years—a very long time—during which the experiments of evolution could lead from the primitive prosimians to the first of the higher primates—the true monkeys and apes.

Our Own Sub-Order, Anthropoidea

THE REST OF THE PRIMATES—the monkeys, apes and man—are included in this group. Its precise origin remains in doubt, but their fossils increasingly appear from the Oligocene onwards. We shall hear more about these later.

This sub-order is comprised of five families, one of man, one of apes and three of monkeys. The five families are divided into those coming from the Old World, the catarrhines, which include ourselves, apes and one of the monkey families. The remaining two monkey families live in tropical America and are the platyrrhines. The terms 'platyrrhine' and 'catarrhine' refer to the disposition of the nostrils, which in the former are placed rather wide apart and in the latter close together. This distinction, however, is not always very clear. In spite of their general resemblances, the two groups exhibit certain contrasting features that suggest an early divergence in their evolution. The more obvious feature that distinguishes the New World monkeys from their Old World cousins is the tendency of the former to develop specialized prehensile tails—a useful adaptation for life in the tropical forests which has provided these animals with a structure akin to a fifth limb allowing them great diversity of movement when brachiating through the tree tops.

New World monkeys also lack the two bare hardened pads found on the buttocks of Old World monkeys. These are often brightly coloured and very prominent in mandrills, baboons, langurs and others. They also have one more premolar tooth on each side of each jaw than the catarrhines, and there are some small differences of the skull, not easily seen, which are important from the point of classification.

FIG. 57. (A) An Old World or catarrhine monkey. (B) A New World or platyrrhine monkey, showing the difference in the disposition of the nostrils. (After illustrations in the British Museum (Natural History), by permission of the Trustees.)

A brief excursion into primate dentition will enable us to appreciate the importance of teeth as primate distinguishing characteristics, and explains why palaeontologists attach such value to fossil teeth.

Regardless of tooth size or shape, all catarrhines (Old World monkeys, apes and man) have the same dental formula. In each half of the upper and lower jaw from front to back in the permanent dentition are found two incisors, a single canine, two premolars and three molars. This is usually written:

$$\frac{2:1:2:3}{2:1:2:3} \times 2 = 32$$

In contrast the *Cebidae*, the family of New World monkeys, have a tooth count of 2 incisors, 1 canine, 3 premolars and 3 molars:

$$\frac{2:1:3:3}{2:1:3:3} \times 2 = 36$$

and the other family of New World monkeys, the *Callithricidae*, which includes marmosets and tamarins of South America, have a tooth count of:

$$\frac{2:1:3:2}{2:1:3:2} \times 2 = 32$$

The dental formula of living or ancestral primates is, however, only one of the clues zoologists can gain from a jaw. The shape, size and spacing of the teeth is equally revealing. Much experience is needed before these clues can be correctly interpreted, and even eminent workers in this field may differ in their interpretation of a single specimen.

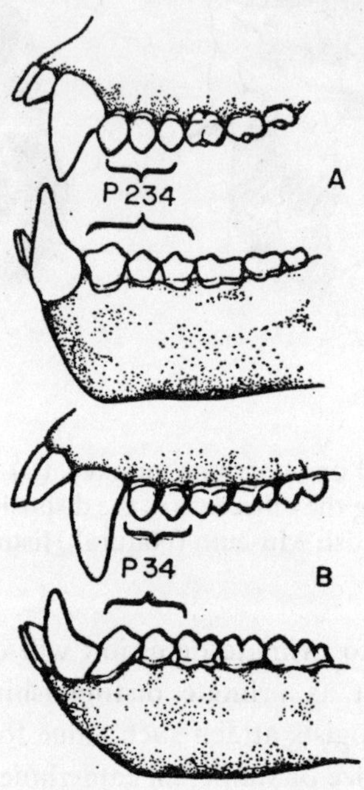

FIG. 58. The dentition of (A) a New World monkey (*Cebus*) and (B) an Old Old World monkey (*Macaca*). (After illustrations in the British Museum (Natural History), by permission of the Trustees.)

The *Anthropoidea* are also divided into three super-families, *Hominoidea*, including man and the anthropoid apes, *Cercopithecoidea*, the Old World monkeys, and *Ceboidea*, the New World monkeys.

Monkeys from the New World

THE PLATYRRHINE MONKEYS are confined to South and Central America, reaching as far north as Southern Mexico. The group includes two families, the *Callithricidae*, the marmosets generally, including the tamarins, and the larger platyrrhines all included in the family *Cebidae*, which numbers such well-known species as the capuchin, the organ grinder's monkey, the woolly monkey (of which a successful breeding colony has been set up in South Cornwall, England) and the larger howler monkeys, that range through their forest territories in bands of a dozen or more. They have a large bony resonating chamber in their throat which facilitates their vocal outpourings, and their roaring is audible for several miles.

FIG. 59. Woolly monkey.

To the marmosets belong the smallest and, in some respects, the most primitive of the extant monkeys. They normally give birth to twins and sometimes triplets—in all other monkeys and higher primates multiple births occur only occasionally. They are largely insectivorous but also feed on fruit, seeds, leaves and shoots. Several marmoset species are known, some as large as a black rat and the smallest being no larger than a mouse lemur weighing about three or four ounces. Particularly attractive are the Golden Lion marmosets, so-called because of the mane of long orange-coloured fur around the face. All marmosets and tamarins retain sharp, curved claws on all the digits except the big toe, which bears a flattened nail. They appear to have lost the opposability of their thumbs. The brain, though relatively large, has a primitive cerebral cortex nearly devoid of convolutions. At one time it was suggested that marmosets were primitive parents of other higher primates, but it is now generally agreed that they went into a kind of evolutionary reverse. Thus their claws seem at one time to have been nails.

FIG. 60. Common marmoset. (After an illustration in the British Museum (Natural History), by permission of the Trustees.)

Monkeys from the Old World

THE CATARRHINE MONKEYS of Africa and Southern Asia all belong to the super-family *Cercopithecoidea*, which consists of only one family, the

Cercopithecidae. There is a rather general tendency in these monkeys towards an increase in size which culminates in the extant apes and, of course, man. None of the catarrhines has a prehensile tail and many are no longer predominantly arboreal in their habits. Indeed the baboons have forsaken the trees altogether to make their homes in unforested regions of Africa.

FIG. 61. Baboon.

Fossil records show that in Pliocene and Pleistocene times some species of macaques were also widespread in Central and Southern Europe. Natural selection has endowed these ground-living species with dangerous temperaments, long snouts and, in the males, prominent canine teeth—a formidable combination of defensive features, especially in animals that live in troops. The Old World monkeys include a great variety of forms. They vary widely in fur and skin colour, some being of most attractive appearance—like the black-and-white *Colobus*, one of the largest monkeys, which grows to three feet in length and has a forty-inch tail. The most distinct and beautiful are the Diana monkeys of the west coast of Africa. They excel in their colour scheme, being black above, pure white below and having pointed white beards and bright red markings on the flanks and hind limbs. There are also the mandrills of West Africa, which resemble the baboons in their body proportions, except that they possess a very short tail. Numerous species belong to the genus *Macaca*, all able climbers but some having become almost entirely terrestrial. Macaques extend their range over vast areas of South-East Asia, from China and Japan to India, and reach into North Africa, where they are found in Algeria and Morocco. The Old World monkeys are four-footed, walking on their flat palms and soles. Probably connected with the common sitting posture is the universal presence in these monkeys of sitting pads—ischial callosities mentioned earlier.

Two other special traits are associated with feeding habits. One group of leaf-eaters possesses saculated stomachs somewhat comparable to those of ruminants, and with the same purpose—the digestion of vegetable foods. The other group has cheek pouches which enable the animals to store food—a most useful adaptation for a creature which has to feed in a hurry and later chews and swallows at leisure when a place of safety has been reached.

Some of the more advanced features of the cercopithecoid monkeys include the expansion of the cerebral cortex of the brain, a greater opposability of the thumb in many genera and more flattened nails on the digits. They also have a more efficient type of placentation and a regular menstrual rhythm.

The Hominoidea—*The Super-Family of Apes and Man*

THE GROUPING TOGETHER in the one category, *Hominoidea*, of man and apes, indicates the close resemblance they show in many anatomical features. The *Hominoidea* are sub-divided into two families: the *Hominidae*, which includes modern and extinct forms of man, and the *Pongidae*, to which belong the modern and extinct anthropoid apes.

Thus living ape species approximate much more closely in their anatomical features to the *Hominidae* than do the monkeys. Apes are similar to man in the configuration and the relative size of the brain; many details of the skull skeleton and dentition; the tendency towards the adoption of an erect posture of the trunk; the absence of a tail and many other features. But apart from structural likenesses there are many fundamental physiological processes in man and apes that are surprisingly similar; for example, the chemical reactions of their blood, the structural details of the tissues through which the young are nourished in the womb before birth, and even the kind of parasitic infestation to which they are susceptible. A good deal of this evidence suggests a genetic relationship, even though it is highly probable that some of the less fundamental resemblances are at least partly due to the result of parallel independent evolution.

Living apes are arboreal in varying degree and have adapted themselves to their particular habits of life by undergoing considerable divergent specialization peculiar to themselves. Evolutionary experimentations have favoured a great lengthening of the arms and a relative shrinkage of the thumb thus allowing the hands to become more hook-like to suit their habit of swinging arm over arm among the branches. The feet are adapted to a grasping function so that apes cannot walk properly on their soles but have to support their weight on the outer side of their feet. They also show little tendency to walk upright on the ground, and while gibbons can occasionally be observed to adopt an upright posture when running along the bough of a tree, this is not true uprightness in the human sense, and they can only accomplish it by bending their knees.

There is a considerable differencc between the Asiatic apes, the orang-utan and the gibbon, and the African species, the chimpanzee and gorilla. The two African species have acquired some terrestrial habits and do not brachiate as freely as their Asiatic cousins, but in the construction of their limbs they also show the same adaptive specialization that is related to an arboreal life.

Gibbons inhabit the dense forests of South-East Asia from Assam and Formosa to Java, Sumatra and Borneo. They are slenderly built creatures of great agility, and they move rapidly through the tree tops swinging along among the higher branches. But, in spite of their skill as arboreal acrobats, gibbons have been known to suffer severe injuries from falls, as is indicated by the frequency of healed fractures observed in gibbons found in the wild. They live in monogamous family groups of five to six individuals comprised of the parents who stay together for life, and any children who have not reached maturity. Their brain volume is quite small, about ninety cubic centimetres, and they are the least intelligent of the higher apes, though due account must be taken of the fact that they are also of much smaller body size, and that brain and body size are related.

In some respects gibbons are more like the Old World monkeys than are the larger anthropoid apes. They have, for example, small ischial callosities and the convolutional pattern of the brain is relatively simple.

A thick and often colourful coat of fur covers their whole bodies, while the other apes (and also man) have only coarse hairs which cover the skin relatively thinly. In all fundamental characteristics, however, such as their dentition, a free vermiform appendix, the absence of tail and other features, they are pongids.

FIG. 62. Gibbon.

The orang-utan, the rarest of the apes, is today confined to the forests of Borneo and Sumatra, but in Pleistocene times it extended its range as far as North China. Little is known about its family life, and there are probably no more than about five thousand individuals alive today. The male orang may reach a height of four to five feet or more, weighing as much as 200 lb.; females are somewhat smaller. Particularly when living in zoos, orang males have a tendency to become very fat, and this emphasizes the large fleshy pads on the sides of the face and the pouch of skin under the jaw.

FIG. 63. Orang-utan.

Unlike the active and agile gibbons, orangs usually swing slowly and deliberately from branch to branch. The ape has weak legs but strong long arms highly specialized for arboreal life. Accordingly, this specialization makes adult orangs very clumsy on the ground. Orangs contrast with the large African apes in the rounded contour of the cranium and in the absence of projecting brow ridges. They are true vegetarians and feed mostly on fruit, being particularly fond of the durian.

The Asian apes have shown increasing adaptation to arboreal life. There is now some fossil evidence to show that by Miocene times some of the primates were already beginning to experiment with the return from the trees to the ground. Three living species are descendants of the ancestral forms in which this trend first appeared. Two of them are our present-day chimpanzee and the gorilla, the third is man. Chimpanzees and gorillas are quite similar to each other in many of their features, and are likely to be closely related. Both show less pronounced adaptations for life in the trees than their Asian cousins the gibbons and orangs, but chimps especially are still essentially tree-dwellers, while the gorilla, especially the mountain

FIG. 64. Gorilla. (After an illustration in the British Museum (Natural History), by permission of the Trustees.)

gorilla from Central Africa, has become a ground-dweller and takes but rarely to the trees. Their habitat also contrasts with their Asian relatives that are strictly jungle creatures of dense equatorial forests. Chimpanzee and gorilla habitat, although also equatorial, is often less dense and permits of freer movement on the ground.

About the same size as the orang-utan, chimp males weigh around 100 lb., whereas gorilla males sometimes reach a height of about six feet and a weight of more than 400 lb.

Active climbers, chimps spend a part of their life in trees but often descend to the ground. They usually proceed on all fours, taking the weight on the soles of their feet and the bent knuckles of the fingers. Frequently they can be seen standing upright but they do not normally walk in this posture. Single families or family groups move about in troops and, like the orang-utan, they also construct crude nests of branches high up in the trees where they spend the night.

FIG. 65. Chimpanzee.

Gorillas usually live in territorial family groups consisting of from ten to fifteen animals, dominated by an old male. Both the African apes are polygamous.

Chimp and gorilla feet are more adapted for walking than those of their Asian cousin the orang, but they are still far removed in their structure from human feet, and lack a pronounced heel—one of the reasons why these apes cannot adopt the human bipedal gait with comfort.

A closer look at ourselves may reveal where we might fit into this picture. To begin with, there is no doubt that we, too, are catarrhines of the Old World. We are hominoids with a similar anatomy to the apes; but, with perhaps the exception of small boys, most of us find tree climbing distinctly cumbersome and our unsuitability as brachiators is clearly reflected in our anatomy. In fact we normally keep our feet solidly on the ground, and our skeleton and anatomy have evolved to suit our bipedal gait and upright posture.

At first glance, hair seems to be absent almost everywhere from our skins, except we have a lot of it growing on our heads, and in the case of males, on our faces. There is also some pubic hair common to both sexes. But in fact, our skins *are* hairy, and the pattern of growth is similar to that found in apes or monkeys, except that in man hairiness seems to have regressed, and most of us are now left with a thin covering which, in blond people, becomes practically invisible. But what made us truly human is our feet. This may seem strange at first, because obviously man's tremendous superiority, which brought him to the pinnacle of the evolutionary tree and made him the most intelligent and diverse creature on earth, is his large brain contained in an inflated head. It is equally probable that man's large brain would never have developed if it had not been for his feet which evolved earlier, allowing some of the early hominoids to walk upright. Because the use of arms and hands was unnecessary for walking they were free to engage in the countless different actions which we associate with human behaviour.

The evidence provided by the living apes leaves a large question mark with regard to the history of locomotion in the evolution of man. We have seen that with the exception of the gorilla they are all at home in the trees and none of them have mastered the bipedal gait. There is some evidence, however, that the more human-like the apes the less are their limbs specialized for the brachiating habit. Perhaps man descended from arboreal acrobats like the gibbons, only later to revert to human proportions. But this seems unlikely. More acceptable is the view that our ancestors were generalized primates at home both on the ground and in the trees. What probably happened was that the tree tops became over-populated with monkeys and apes, whose remaining descendants are the living species of today, leaving the terrestrial environment open to more adaptable anthropoids.

The process of evolution attempts to place creatures in all available niches in a given environment, and thus natural selection made most of the pongids into skilful brachiators, who were still more at home in the tree tops than on the ground. This left a niche for entirely ground-living primates which was occupied by the hominids.

What are our true relationships with living primates, and are there any connecting links among fossil remains which can shed some light on this question? Before an attempt is made to answer the question, let us make it clear from the start that no living ape can possibly have been our ancestor, and any comparison made earlier between living primates and man was only meant to shed light on stages of development through which the ancestry of man could have passed. The direct ancestors of living apes were earlier ape species now extinct. Similarly, man's ancestors were primitive men and as we shall see, ape-men and their ancestors. Clearly, in far distant Eocene times tree shrews must have lived that gave rise to the evolutionary line of pongids and hominids. But these two families left each other's ancestral company a very long time ago. In fact, on present evidence, it is likely that the hominids, of which man is the only representative living today, began their separate lines of evolution as long as 25 to 30 million years ago.

First Known Fossil Apes

We cannot say for certain where the first step in hominoid evolution took place. What we do know, is that as long ago as 35 to 40 million years, during the Oligocene, primitive primates must have lived in what is today Egypt and Burma. Here they have left some remains of their teeth and jaws as fossil evidence. Admittedly the remains are scanty and poorly preserved but they do point directly at a number of evolutionary radiations which led to the higher primates, especially the apes. These creatures seem to have lived during the time when the majority of the prosimians (lower primates) were disappearing from the world stage, to be followed by more well-defined lines of monkeys and apes leading to those living today.

Particularly important are two lower jaws that were dug out of the Oligocene deposits in the Fayum depression in Egypt. In 1911 Dr Schlosser described these fossils in detail, naming them *Parapithecus* and *Propliopithecus*. A great deal of justifiable importance has been attached to these two lower jaws, for they belong to the earliest higher primates known. Coming from the Oligocene deposits, they are probably somewhere between 35 million and 40 million years old.

Skulls and parts of skulls, such as jaw bones and teeth, are among the most valuable fossils, for they supply a wealth of information. For example, a primate's skull can indicate the size of the head, and hence give some indication of the size of the whole animal. It also gives information about the size of the brain, indicating whether the primate was primitive or advanced. Further, the shape of the base of the skull is indicative as to how the head sat on the spine, and whether the animal was bipedal or quadrupedal. Important, too, as mentioned earlier, are fossil teeth. Made from calcified material and capped with enamel, they are usually well preserved, and their shape and number, as well as the dental arch into which they are fitted, reveal many important characters and habits of their owners. It immediately became apparent that *Propliopithecus* had a catarrhine tooth count, and thus belonged to the *Catarrhinae* (Old World monkeys). It may in fact have been a primitive hominoid, on one of the lower rungs of the hominoid ladder.

Parapithecus is a somewhat more doubtful character. It was about the size of a little squirrel monkey we see today, and has been hailed by some as a generalized ancestor of the hominoids. It seems, however, that its teeth are letting it down in this respect. Indeed, because of the generalized and undifferentiated form of the front teeth, there has been considerable doubt whether it had already acquired, in the lower dentition, the dental formula characteristic of the modern Old World monkeys and anthropoid apes. Recently a number of experts have favoured the idea that these animals were primitive herbivores, which have been shown to have similar teeth.

Dr Elwyn Simons of Yale University has been successful in finding further fossils to support the earlier Egyptian finds. Some of these represent new species of *Parapithecus* and *Propliopithecus*; others he discovered were entirely novel types of hominoids differing from each other and distinct from *Parapithecus* and *Propliopithecus*. Thus fossils of hominoids which lived some time during the beginning and later part of the Oligocene are fairly plentiful. Still small in size and primitive in traits, these creatures were somewhere near the bottom of the hominoid evolutionary ladder.

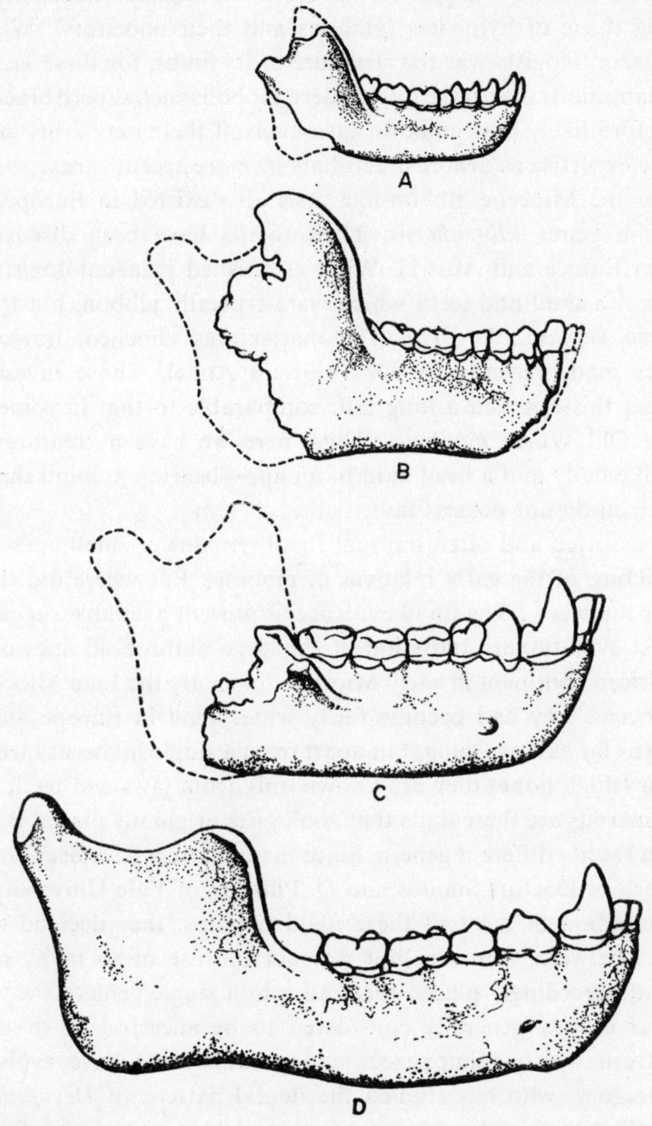

FIG. 66. The mandibles of (A) *Parapithecus*, (B) *Propliopithecus*, (C) *Pliopithecus*, (D) modern gibbon. (After illustrations in the British Museum (Natural History), by permission of the Trustees.)

Later, in the early part of the Miocene period, probably about 25 million years ago, a great many apes existed in East Africa, which are known not only from jaws and teeth but also from portions of the skull. The smallest of these is a gibbon-like creature, *Limnopithecus*, with teeth resembling those of hylobates (gibbons and their ancestors). What surprised palaeontologists was the structure of its limbs, for these lacked the special adaptations that make our modern gibbons such superb brachiators. It is therefore likely that gibbons have evolved their long arms and with them their expertise as arboreal acrobats in more recent times.

During the Miocene gibbon-like apes also existed in Europe, represented by a genus *Pliopithecus*. Their fossils have been discovered in deposits in France and Austria. What astonished palaeontologists about this ape was a skull and teeth which were typically gibbon, but the hand bones were very monkey-like. The matter was clinched, however, by discoveries made recently by Dr Friderun Ankel, whose investigation proved that this ape had a long tail, comparable to that in some of the tree-living Old World monkeys. Thus, here we have a creature with a monkey-like body and a head akin to an ape—bearing in mind that living apes and men do not possess tails.

These assorted and often unusual fossil remains of small apes provide a good picture of the early relatives of gibbons. But where did the large apes enter the scene? The fossil evidence at present available suggests that the earliest evolutionary radiation of the large anthropoid apes occurred on the African continent in early Miocene times. By the Late Miocene and Early Pliocene they had become fairly widespread in Europe and Asia, extending as far east as China, but apart from a single humerus (arm bone) and femur (thigh bone) they are known only from jaws and teeth.

So numerous are these finds that zoologists originally placed them into more than twenty different genera. Since then, matters have been simplified by the work of Doctors Simons and D. Pilbeam of Yale University. After a re-examination of most of these fossil remains, they decided that the differences between them did not warrant the use of so many different genera and accordingly placed them all into a single genus *Dryopithecus*. This genus is now generally considered to be ancestral to the kind of animals from which chimpanzees and gorillas must have evolved. Dr W. K. Gregory, who has studied the dental pattern of *Dryopithecus* in great detail, has found evidence to suggest that the teeth of gorillas, chimpanzees, orang-utans and man could have all evolved from the generalized dentition of *Dryopithecus*. This incidentally is one reason why the apes and man are all grouped together into the super-family *Hominoidea.* In other respects, however, *Dryopithecus* appears more like the African apes. These ape ancestors also reveal a tendency towards a projecting mouth, an evolutionary adaptation which facilitates the eating of coarse vegetation and fruit. It has become particularly prominent in the gorilla.

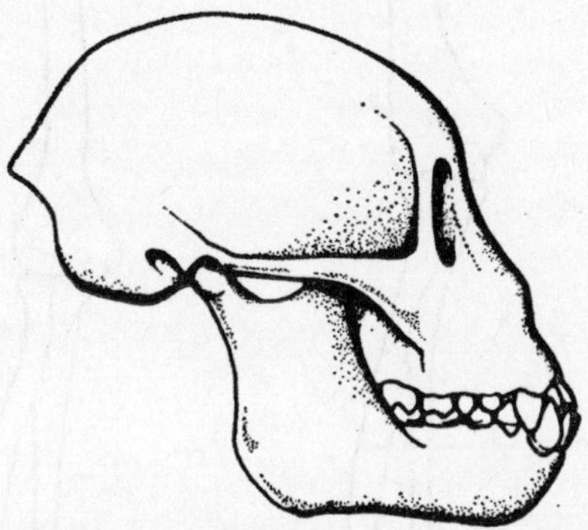

FIG. 67. The skull of an East African Miocene ape, *Proconsul africanus*. (After an illustration in the British Museum (Natural History), by permission of the Trustees.)

We now come to the important fossil *Proconsul.* It left an abundant supply of remains in Lower Miocene lake deposits by Lake Victoria in Kenya. Dr Louis Leakey and his associates have recovered most of them. Here, at last, we have complete skeletons which represent three species of different sizes corresponding to a gorilla, a large baboon and a chimpanzee. *Proconsul* received its name from Dr Hopwood of the British Museum, after he had examined the first of the remains and thought it to be a forerunner of the chimpanzee. At that time London Zoo possessed a chimp called Consul, and so Dr Hopwood decided to name his fossil after him. *Proconsul* may be regarded as a sub-genus of *Dryopithecus* discussed earlier. It was definitely an ape, but what makes this fossil particularly exciting is that it resembles a common ape-man ancestor more than did any later fossil ape. The skull and jaws are definitely of an advanced ape type, but the low, rather oblong, eye sockets, the fore-shortened muzzle and the smooth forehead are in many respects unape-like and tend toward the human. *Proconsul* could have been the ancestor of the modern chimpanzee. Furthermore, its structure is so generalized that it could have been ancestral to the gorilla, or even man. An active and agile creature, its skeleton reveals that it was of slender build, and the few arm bones that

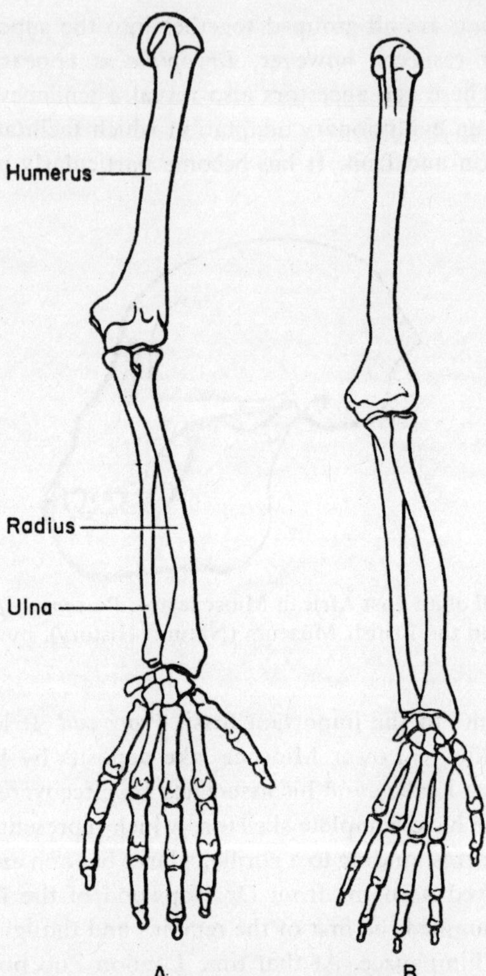

Fig. 68. The fore-limb skeleton of a chimpanzee (A) and *Proconsul africanus* (B) (not drawn to the same scale). Note the delicate construction of the limb bones in the fossil ape, the shorter fore arm relatively to the upper arm, and the longer thumb. (After illustrations in the British Museum (National History), by permission of the Trustees.)

have been discovered indicate that it could not have been as clever an arboreal acrobat as the modern gibbons. *Proconsul* probably spent a good deal of its time on all fours on the ground and unlike its descendants, the modern apes, lived in open savanna country where woodlands and prairies formed its habitat.

Another fossil primate of great interest is *Oreopithecus*. Known since the 1860s, its fossil remains have been recovered from lignite mines near Monte Bamboli in Italy. At first only lower and upper jaws were discovered showing a dentition that placed it somewhere among the Old World monkeys. It was only after Dr Johannes Hürzeler of the Natural History Museum of Basle had re-examined the teeth and jaw fragments that it became apparent that earlier drawings and descriptions of these fossils were quite inaccurate.

Hürzeler now set about enthusiastically to see whether more fossil remains of this peculiar primate could be discovered. He was lucky. Eventually, in 1958, after some systematic excavations at the working face of a deep underground mine, the distorted but almost entire skeleton of an *Oreopithecus* was exposed. Further finds indicated that *Oreopithecus* had not been a particularly rare creature of his time in this area; and, after studying its skeleton, Hürzeler came to the conclusion that this Southern European primate was definitely not a monkey. More surprisingly, both skull and jaws indicated that it might not only have been a hominoid but a hominid, thus belonging to man's own family the *Hominidae*. Unfortunately the teeth did not fit into the picture. They had certain characteristics which are not found in any of the hominids. This in the end induced palaeontologists to remove it from close association with either the pongids or hominids, and to grant it the status of a separate family, *Oreopithecidae*.

Oreopithecus is of particular interest because, in the later part of the Tertiary era, we know so little about possible pongid or hominid ancestors, and there is a gap of about 25 million years between *Proconsul* and *Australopithecus*, whom we shall meet shortly.

Traces of a true giant of the past, even larger than our present-day gorillas, were discovered by Professor G. H. R. von Koenigswald in a Hong Kong drugstore. There in 1935 the Professor purchased a number of huge primate molars sold to him as 'Dragons' Teeth'. A good deal of excitement surrounded this discovery, and some even thought that the teeth might have belonged to an early man. The odd thing is that *Gigantopithecus* lived in the middle of the Ice Age about half a million years ago after all the fossil apes described earlier had probably perished.

When, in 1956, more of its fossil teeth and also three lower jaw bones had been discovered in various locations by Chinese scientists, *Gigantopithecus* was placed fairly and squarely into the family of apes, and has remained there ever since. Doubtless a ground-dweller of hefty proportions, it must have been a hardy creature for it lived in the inclement condition of the Ice Age and had man—who at that time was already much in evidence—to contend with as a competitor.*

The Man-Apes of Africa

FOSSILIZED plants and animal bones are not a great rarity. An amateur palaeontologist can assemble a handsome collection providing he has a trained eye and knows where to look for them. It is, however, quite another matter to discover some of the most important fossils relating to man's ancestry. Skill as well as a great deal of luck are required. Professor Raymond Dart and Dr Robert Broom were blessed with both.

The scene was set in 1924 at a lime works near Taung, in South Africa. There, during a blasting operation, a miner came across a small fossil fragment representing the brain cavity of what resembled a chimpanzee. Together with other fossil fragments it was sent to Dr Raymond A. Dart, Professor of Anatomy at Witwatersrand University at Johannesburg. Soon more bones belonging to the skull were discovered at the site, until eventually the face, jaws and teeth could be assembled to yield the fossil skull of what appeared to be a young ape, perhaps six years old.

Dart soon realized that here was a skull that had not belonged to an ordinary ape. Admittedly, the muzzle was well marked and rather ape-like, and the brain case was small. But the absence of brow ridges, the shape of the dental arch and the small canines looked remarkably human.

Dart was particularly impressed by the roundness of the skull which suggested that the creature might have walked upright. To this specimen he gave the name *Australopithecus* (Ape of the South). In his notes he mentioned that the specimen represented an extinct race of apes intermediate between living anthropoids and man.

When his report was published in the English scientific journal *Nature*, the reactions from his fellow scientists were generally unfavourable and critical. Most of them dismissed the idea as a figment of imagination, thought up by somebody who lacked experience in this particular field. Nearly all the leading anthropologists, with the exception of Dr Broom,

* In 1958 a joint Yale University–Punjab University expedition discovered a *Gigantopithecus* jawbone in the Siwalik Hills of India, where it was found in sediment laid down between 5 and 9 million years ago.

believed that Dart had committed a serious blunder and that the little Taung skull simply represented an extinct variety of chimpanzee.

For a time interest continued to centre around the limestone works at Taung but no further fossils connected with the controversial skull were discovered. Indeed years went by without further developments, but during this time Dart did not remain idle. He had found a staunch supporter in the person of Dr Robert Broom. A fossil expert, Broom had studied at Glasgow University where he took a doctor's degree in 1889. After receiving reports of strange reptile fossils in South Africa Broom decided that this was the country for him and accordingly he set sail for South Africa to arrive in Capetown in January 1897.

He set up his medical practice near the fossil-rich Karroo country, where after years of work his researches on reptilian fossils enabled him to successfully bridge the gap between mammals and reptiles. He was able to demonstrate that the former had actually evolved from the latter.

Broom had become interested in the Taung discoveries and together with Dart he continued to study the Taung skull. The more they looked and measured the more convinced they became that Dart's original verdict had been right.

Particularly significant to Broom was the fact that the large opening leading to the brain (*foramen magnum*) was situated farther forward than in apes, indicating that this little creature must have walked erect, like us. Dr Broom's problem was one of time. He had to make a living as a medical practitioner which allowed him too little time to pursue his palaeontological studies. Fortunately General (later Field-Marshal) Smuts took an active interest in the pre-history of South Africa. He persuaded Broom to give up his medical practice, offering him a post as curator of vertebrate palaeontology and physical anthropology in the Transvaal Museum in Pretoria. There for over a year he continued with his studies on fossil reptiles and it was not until 1936 that he began to devote his time more fully to the search for further evidence of *Australopithecus*.

Limestone caves had proved fossil treasure houses, and Broom continued to search among their debris to find remains of sabre-toothed tigers, giant baboons, rats and other mammals. His findings, freely published in the local press, came to the attention of students of Dart's, who informed him about some small baboon skulls they had found in a limestone works at Sterkfontein near Johannesburg. Broom followed up this lead immediately and one Sunday morning in 1936 he was driven out there to investigate. At first he and his students visited a quarry where for many years a company had been blasting lime for its kilns. Broom soon learnt that the manager, Mr G. W. Barlow, had at one time worked at Taung, knew about the man-ape fossil skull discovered there, and whenever the opportunity arose was doing a brisk tourist trade in fossilized bones. Mr Barlow promised to look out for further evidence of fossil skulls similar to that found at Taung. The 17th August 1936 proved to be one of Dr Broom's lucky days, for on that Monday morning Barlow presented him with a fine fossil brain cast in the form of a stone that bore the exact impression of a skull that had once lain there. Broom immediately recognized it as the brain cast of a man-ape.

There now began a feverish search through the debris, and during that day and the next Broom himself discovered the cast of the top of the head, all of the base of the skull, some pieces of the upper jaw and actual fragments of the brain-case.

Thus, within three months of starting work, Broom was in possession of an almost complete adult skull of *Australopithecus*. During the next two years the doctor paid many more visits to the same site, recovering more fossils, including the knee end of a thigh bone and some isolated teeth and skull fragments. Then in June 1938 came another startling discovery. When Broom arrived at the site on that memorable day Barlow met him with: 'I have something nice for you this morning.' It was the upper jaw of a large man-ape with one tooth in place. It soon became clear that this fossil had not come from Sterkfontein at all; indeed, it had been given to Barlow by a fifteen-year-old schoolboy, Gert Terblanche, who lived on a farm at Kromdraai, about two miles away.

Arriving at the boy's home, Broom found that he was at school, but the family were able to show him the site on a hill where Gert had dug the jaw fragment out of a deposit of bone breccia.

Broom then hurried to the school where he questioned Gert about his fossil finds. The doctor was thrilled to discover that the fossil jaw was by no means the only item the lad had recovered, for Gert also produced four beautiful teeth of incalculable value to anthropologists. This was almost too good to be true. Taking the boy with him, Broom set off for the hill at Kromdraai where Gert had excavated the fossils, and there to Broom's delight Gert presented him with a piece of lower jaw containing some fine teeth. Further excavations during the next few days yielded a part of the face and the greater part of the left side of the skull.

Broom noticed important differences between this Kromdraai skull and that of *Australopithecus* found at Sterkfontein. The jaw was more powerful, the teeth larger, and the face appeared flatter. This tended to make the fossil appear more ape-like. Yet the form of the teeth emphasized human qualities. In any case the differences between this ape-man and earlier finds were important enough to place it in a different genus, named by Broom *Paranthropus robustus*.

Clearly evidence was gradually accumulating that Dart had been right. Man-apes must have lived in that part of Africa and probably elsewhere, and there was no longer any denying that a new chapter in anthropology was about to be written.

Field-Marshal Smuts, who had always taken an active interest in Broom's work, offered him financial support to continue with his researches after the war, and he was so successful at this that by the time he died in 1951, at the age of eighty-four, fossil parts of nearly thirty different specimens of man-apes had been recovered, and palaeontologists the world over no longer doubted that these fossils belonged to upright-walking, ape-like creatures possibly directly related to man.

Olduvai Gorge, in Northern Tanzania, is a miniature Grand Canyon sliced through three hundred feet of sediment covering the bed of a prehistoric lake. It has become one of the world's most famous archaeological sites largely because of the dedicated efforts of Dr L. S. B. Leakey and his wife, Mary.

The existence of the gorge had not been known until 1911, when a German entomologist nearly tumbled into its depth while catching butterflies on the plateau above. The gorge proved to be a veritable storehouse of pre-history.

Dr Leakey had been making a systematic study of East African pre-history since the mid 1920s when, in 1931, he and his family began work at Olduvai.

He was born in Africa in 1903 where his parents were missionaries among the Kikuyu tribe. At the age of sixteen he left for England to attend school and his later studies at Cambridge included the Stone Age history of Kenya. He returned to Africa in 1924 on an archaeological expedition, to be soon afterwards appointed as curator to the Coryndon Museum, Nairobi.

Numerous reports and articles of his work at Olduvai have been published in the *National Geographic* magazine. They tell of his truly remarkable successes in finding fossils, these being largely due to patience, experience and hard work.

Olduvai Gorge turned out to be an anthropologist's paradise. The walls have been neatly divided into five distinct beds, the lowest of which was laid down between 1·9 and 1·5 million years ago. The 17th July 1959 marked another memorable day in the annals of anthropology. Dr Leakey had been taken ill with a fever and had to remain in camp, but in order that there should be no delay with their work his wife, Mary, set off on her own to the excavation site. Working her way up a hillside, she recognized a piece of skull lodged in the ground, and at some distance from it two large teeth. Her long experience immediately told her that these bone fossils were not those of an ordinary animal. It seemed that at long last their twenty-eight year long search was to be rewarded with a fossil find of early man. There was added interest in the fact that the fossils were situated in Bed I and must therefore have been of considerable antiquity. They named it *Zinjanthropus boisei* after the old Arabic word 'Zinj' for East Africa, and *boisei* in honour of Charles Boise, whose financial backing had assisted the Leakeys to maintain their researches over many years.

Five years later, in 1964, a lower jaw was discovered fifty miles away from the *Zinjanthropus* find, at Peninj on the shore of Lake Natron. The jaw matched the skull reasonably well and is thought to have belonged to the same type of creature.

The skull of *Zinjanthropus* was first estimated to be about six hundred thousand years old. A surprise awaited anthropologists when chemists applied the potassium-argon test to volcanic ash from above and below the find in Bed I. It revealed an age of one million, seven hundred and fifty thousand years. The striking fact here is that the jaw subsequently discovered in 1964 is at least a million years younger than the skull.

Even older fossil bones, believed to have belonged to an *Australopithecus*, were discovered in 1965 by Dr B. Patterson in an early Pleistocene deposit near the southern end of Lake Rudolf, Kenya. It was an arm bone of a hominoid, similar to that of a chimpanzee. But this part of chimp anatomy is very similar to that of humans, and the finding of antelope remains at the same site makes it unlikely that the region was forested, and hence the home of chimps. Much more likely this creature belonged to a species of *Australopithecus* or *Paranthropus* or an allied form of ape-man. The fauna of this deposit has been estimated close to 2·5 million years, considerably older than the Olduvai Gorge deposits. These discoveries then take the man-apes back to the time near the beginning of the Pleistocene, now estimated to have started about 3 million years ago.*

* The latest finds of *Australopithecus* fossils were made in the Omo River valley, Ethiopia, in 1967–8. Teeth discovered there belonging to *A. africanus* are about 3 million years old.

A review and evaluation of the man-ape evidence made it clear that the australopithecines of those early days differed slightly from each other. Accordingly they were divided into three species, *Australopithecus africanus* from Taung, Sterkfontein and Makapansgat. Another species was recognized from Kromdraai and Swartkrans and named *robustus*. That discovered at Olduvai (*Zinjanthropus*) was named *boisei*. Some anthropologists have maintained that *Australopithecus africanus* is a direct ancestor of man. Certainly there doesn't seem to be anything in his features that could not have evolved in our direction, and his skull and teeth bear considerable resemblance to the genus *Homo*.

Paranthropus, the large gorilla-like ape-man from Swartkrans, seems to have been a different creature altogether. Dr J. T. Robinson has argued that they have evolved from one of the australopithecines and later became extinct. Evolutionary pressure towards an increase in body size can be explained when one considers the number of serious potential enemies *Australopithecus*, who grew to only about four feet and weighed no more than 80 lb., must have had. A 150-lb. ape must have found it easier to deal with its natural enemies, particularly the carnivores, that were abundant in its day. Most of its teeth were larger than those of *Australopithecus*, this being a result of the increase in size of the whole organism. Dr Leakey's *Zinjanthropus* from Olduvai Gorge has, after further consideration, also been classified as an australopithecine of the *Paranthropus* type. It probably proceeded in a similar manner to our modern gorillas and it is certain that it never evolved the relatively efficient bipedal gait of *Australopithecus*. The two species must have appeared as different from each other as chimps and gorillas do today. *Australopithecus*, like ourselves, was probably omnivorous, and did not, like most of the apes, restrict its diet to fruits and vegetation.

Teeth are a good if not entirely accurate pointer to a creature's eating habits, and those of *Australopithecus*, like our own, strongly suggested a mixed diet of meat and vegetables. Dart believed that the many animal bones found in its caves could not all have been carried there by carnivores, such as the hyaenas. There were, for example, many smashed baboon skulls of doubtful origin. Dart surmised that the man-apes used larger bones to slay their victims, perhaps because they enjoyed eating their brains, or simply to rid themselves of what must have been a very numerous competitor in their environment. It is reasonable to speculate that other animals also fell victim to their hunting expeditions.

On the other hand, *Paranthropus*, the hominid gorilla, was a vegetarian.

Evidence derived from the teeth revealed this. The tiny flakes chipped from the top of the crowns down the side of the enamel did not, according to Dr J. T. Robinson, result from crunching resilient material like the bones of animals. They were caused by the presence of grit in the diet, suggesting that the plant food included roots and bulbs.

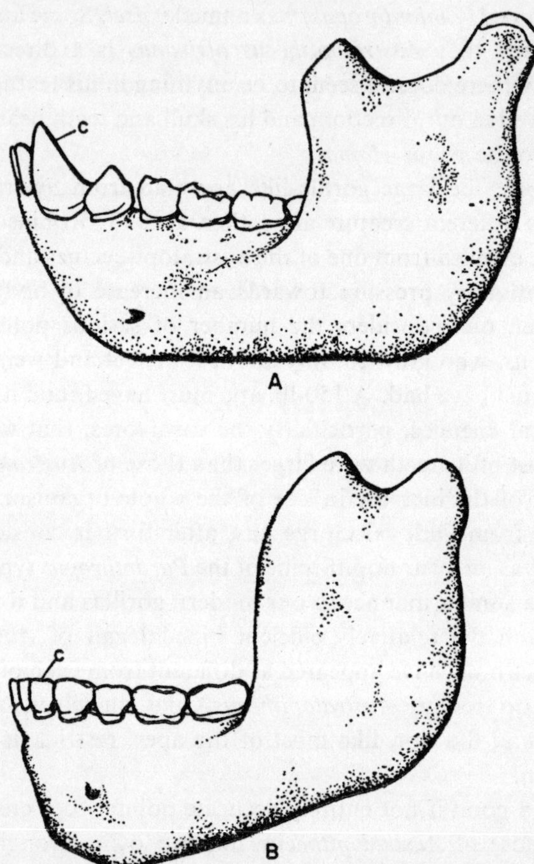

FIG. 69. The lower jaw of (A) an orang-utan, and (B) one of the australopithecines. Note that the orang-utan canine (c) is very different from that of a man-ape. (After illustrations in the British Museum (Natural History), by permission of the Trustees.)

If on the other hand we accept that the forest-dwelling *Paranthropus* represents the original australopithecine stock and *Australopithecus* evolved from them, it is intriguing to speculate why this evolutionary diversion took place at all.

Evidence comes from the Kalahari sands of Southern Africa to suggest that the Miocene was a period of expanding forests in these areas, but that during the late Miocene and early Pliocene there were long periods of drought with accompanying shrinking forests. Thus suitable habitats for the original vegetarian, *Paranthropus*, and for other animals requiring a forest habitat, became increasingly scarce through the late Tertiary. This must have resulted in severe competition among them for feeding and breeding places. As the forests shrank, the grasslands and other more arid environments expanded, providing additional living space for those animals adapted to, or capable of adapting to, such conditions. These climatic changes, of course, occurred very gradually and the australopithecines living in areas where severe drought became a frequent occurrence must have found plant foods increasingly difficult to locate, and so some of them began to experiment with a diet of meat.

It is noteworthy that among the living apes a number of them will readily eat meat. Gorillas in zoos have had their diets successfully supplemented by meat, although there is no evidence that they eat meat in the wild. On the other hand, studies by Dr J. van Lawick Goodall have proved conclusively that chimpanzees in the wild kill and eat the occasional small monkey. Thus a gradual change from a vegetarian diet to meat would not have proved an insurmountable obstacle to some of the australopithecines, leading to the evolutionary diversion which led to the omnivorous *Australopithecus*.

Tool-using today is by no means confined to primates, and it is highly probable that primates of the degree of development of australopithecines, which walked erect on their hind limbs and thus had emancipated front limbs, sometimes used tools. As mentioned earlier, Dart had investigated this possibility, arriving at the conclusion that the man-apes selected the teeth and bones of dead animals, such as antelope jaws lined with teeth, for simple scraping and cutting operations, and the heavier limb bones as defensive or offensive weapons. He also drew attention to certain fossil bones found on the man-ape sites that appeared to be polished from use.

Recent observations by Dr J. van Lawick Goodall on wild chimpanzees in the Gombe Stream Reserve have provided fascinating data on their ability to use tools. She observed for example that leaves were frequently used as drinking cups and for wiping various parts of the body. Chimps also used sticks or grasses as 'investigation probes' and at an artificial feeding area they were seen to use sticks for prising open boxes containing bananas. However, one of the most exciting examples of tool-using concerns an infant chimp that attempted to hit an insect on the ground with a stick.

FIG. 70. A restoration of the appearance of an *Australopithecus*. (From an illustration in the British Museum (Natural History), by permission of the Trustees.)

Chimps are the most intelligent living apes, but their brains are smaller than those of the australopithecines, so that the possibility remains that the latter not only used bone as tools but also made and used primitive stone tools, although to date no definite evidence of this exists.

Charles Darwin had already suggested that the reduction in canine tooth size may have resulted from the use of tools. Effective tool-using could only have become possible after erect posture had been acquired, and the key to the origin of the australopithecines lies in the change to standing upright and efficiently walking and running on the hind limbs. This represented a major evolutionary experiment which opened up entirely new possibilities for these primates. Erect posture clearly gave rise to tool-using. This, in turn, may have led to the reduction of canine teeth and most important of all to the expansion of the brain. J. T. Robinson believes that the earliest australopithecines must have originated considerably before the Pleistocene period, possibly even as early as the Miocene, somewhere between 25 million and 11 million years ago.

The important thing about all this is that we now have definite evidence that, in the early Pleistocene, there existed on earth upright ground-walking creatures, with brains not much larger than those of present-day chimpanzees, and possessing a pair of hands which allowed them a freedom of action far surpassing that of any other living species on earth at the time. Had the evolutionary experiments failed to produce erect, ground-walking animals, capable of using their fore limbs for activities other than simple body support, it is quite possible that man might never have evolved into his present form.

The discovery of *Australopithecus* also indicates that man did not emerge from the fossil creature *Dryopithecus*, for this reveals very definite adaptation trends to the apes with their specialized teeth. *Australopithecus* was an erect-walking hominoid with hominid teeth that show some primitive aspects compared to our own, but are certainly not ape-like.

Fossil evidence today abundantly shows that many millions of years ago the hominid and pongid lines of development had taken their separate evolutionary paths and it has been the ambition of anthropologists to discover the period in prehistory when they might have diverged. Australopithecines can now be traced back 3 million years. They had a well-formed hominid dentition and a skeleton enabling them to walk upright, indicating that their evolution started a long time ago. The distinctly ape-like dryopithecines furthermore suggest that they might have been a separate evolutionary line in the Miocene or even earlier.

This now takes the period in time when the hominid and pongid evolutionary lines separated some 25 to 30 million years ago. Fortunately, fossil finds have revealed that the gibbons had begun to evolve separately considerably earlier than this, so that even 30 million years, as an approximate date when our own ancestral lines parted from those of the pongids, is not unreasonable. Factors which count rather against this very early separation of the pongid and hominid ancestral lines are certain similarities in blood chemistry found in the chimpanzee, gorilla and man. These, however, may be cases of parallel evolution where two already distinct lines evolve similar features through natural selection.

Whatever else may have happened at the beginning of the Miocene, the hominid and pongid evolutionary lines appear to have parted company when the pongids were already evolving their specialized features for a life in the trees and front teeth adapted to a diet of fruit and vegetation. On the other hand the hominids showed a strong tendency to remain on the ground to become omnivorous feeders evidenced by a more generalized dentition.

It is also possible that the ancestors of such excellent arboreal acrobats as the gibbon and the chimp spent much of their time on the ground and only later returned to the trees to evolve into the brachiators they are today. An analogy to this situation might be found among sea-living mammals, the whales and porpoises, whose ancestors left the ocean as primitive amphibians only to return later at the mammalian stage of their evolution. Here, of course, the time lapse involved is much greater, probably ten to fifteen times as great. In terms of structural changes, it meant the evolution of an early amphibian into a mammal, passing somewhere on the way through the reptile stage. These changes, therefore, are much more fundamental, but they do illustrate that the experiments of evolution can back-track as far as the environment is concerned. All this, however, is mostly speculation. Nobody as yet really knows how our present-day pongids lived in prehistoric times.

A Possible First in the Hominid Line

DR E. SIMONS of Yale University did more than anyone to clarify the somewhat involved story of *Ramapithecus*'s jaws. When Dr G. E. Lewis of Yale studied fragments of the upper jaw of an ape discovered in 1932 he found that the face projected only a little to the front. He named it *Ramapithecus brevirostris*. It had distinct hominid characteristics, including a small canine tooth. At the time, unfortunately, an ape's lower jaw (showing a premolar tooth resembling a canine, a distinct ape characteristic) was falsely assigned to it. Worse still, when these fossils were dated they appeared to come from the late Pliocene, which would have left an insufficient time span of only a million years for such an ancestor to evolve into an australopithecine. In 1961, at Fort Ternan in Kenya, Dr Leakey had discovered other jaw pieces very similar to the original *Ramapithecus* upper jaw. This *Kenyapithecus* was found to be of a much greater age, which placed it into the Lower Pliocene, 11 million years ago. In view of this new information, Dr G. E. Lewis re-dated the original fossil specimen, only to find that it, too, belonged to the Lower Pliocene. Eventually, Dr Simons completed the picture by including *Kenyapithecus*, and a number of other specimens that had been assigned different names, in the same group, demonstrating conclusively that primates belonging to the genus *Ramapithecus* were not apes (*Pongidae*) but probably some of the earliest members of our own family, the *Hominidae*, from the early Pliocene. Unfortunately only jaw pieces became available, so we know nothing of the animal's posture; but its front teeth and the short face suggest an animal where the hands were even more important for conveying food to the mouth than they are in today's African apes.

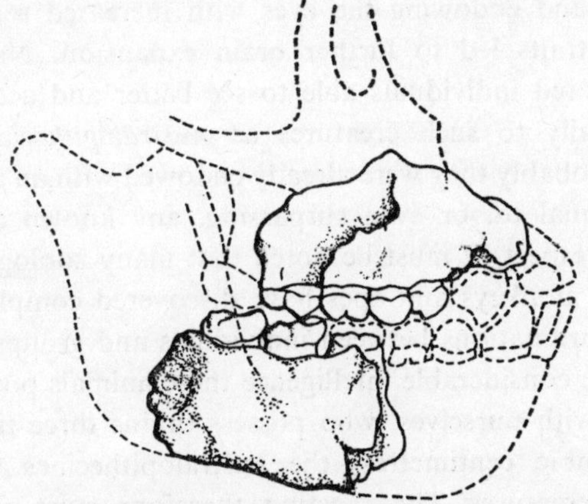

FIG. 71. The jaws and face region of the Pliocene ape *Ramapithecus*. A possible first hominid. (From an illustration in the British Museum (National History), by permission of the Trustees.)

The Developing Brain

Ramapithecus and *Australopithecus* had man-like teeth. The latter even walked upright, but neither of these creatures possessed a brain approaching ours in size and complexity. Skull casts of the australopithecines show an average cubic capacity only slightly larger than 500 cubic centimetres, which is the average brain size for an adult gorilla. It is noteworthy that *Australopithecus* was considerably smaller and lighter than living gorillas, and since brain size is normally related to body size the former had a relatively larger brain than the latter.

A source of considerable speculation and argument among anthropologists is how this remarkable organ, the brain, developed to its present size in man over the relatively brief evolutionary period of the last 2 million years.

The answer to early brain development in primates can be found in Professor Wood Jones's arboreal theory. Once the evolutionary experiments had endowed primates with stereoscopic vision and true hands which enabled them to carry out much more efficient operations than were possible by the simple use of front paws, jaws and teeth, the brain responded by increasing in complexity and size allowing the hands to cope with more complex tasks and endowing the eyes with increased sensitivity. These newly formed traits led to further brain expansion. Natural selection obviously favoured individuals able to see better and accomplish more, leading eventually to such creatures as *Australopithecus* manipulating simple tools. Probably they were already endowed with an advanced social organization equal to, or even surpassing, any known among animals today. In this respect, it must be noted that many zoologists engaged in field studies of monkeys and apes have discovered complex social interactions and coordinations between individuals and groups, expressing to some degree the considerable intelligence these animals possess.

Compared with ourselves, who possess brains three times as large—about 1,450 cubic centimetres—the australopithecines were still very under-brained creatures. The question, therefore, must now be asked—how came we to be the possessors of our enormous brains which made us human?

The process of further rapid brain expansion may have started with a species of man-ape completely able to use bipedal gait, and possessing hands already partly skilled for the shaping and use of simple bone and stone tools. These accomplishments doubtless made him a more efficient hunter, and it is likely that such a creature would also have been able to employ a system of communication akin to our speech, possibly preceded and accompanied by hand signals. This would, of course, further enhance his ability to live successfully in groups, and to outwit the creatures that shared the same habitat. From this stage on, any evolutionary experiments leading to further enlargement and improvements of the brain must have met with immediate success, leading in turn to added manual and social skills.

The Appearance of the First Men and their Environment

TODAY, all hominids that evolved further than the australopithecines are grouped generally in the genus *Homo*. The processes of evolution are normally so very gradual that it is impossible to say exactly when the man-apes became more like man. Evidence of fossils, however, indicates that men more like ourselves than apes had lived in various parts of the Old World in the later part of the Pleistocene epoch, about 700,000 to 1 million years ago, when the man-apes were still much in evidence. These men were characterized by a brain considerably larger than that of the man-apes, and their skeletons had become very similar to those of our own.

Although only tentative evidence exists to suggest that the man-apes were able to use natural objects as tools, early men certainly made and used stone tools. Probably they had considerable speaking ability, too.

These men did not suddenly arrive on the scene. Their beginnings can now be traced back some 1 million years. In fact, our own species, *Homo sapiens*, is the most recently developed kind of man, where the special features of the hominids have been carried to their fullest development. There were many kinds of men before us and a good deal is known about them today. Their more rapid evolutionary advance seems roughly to coincide with the advent of unstable and colder world climates.

During much of the 65 to 70 million years of the Tertiary period tropical climates extended to much higher latitudes than at present. Climatic conditions on earth were generally warm and stable. With the approach of the Pleistocene, however (the name is derived from the Greek *pleistos*, most, and *kainos*, recent), about 3½ to 4 million years ago, cool and temperate climates began to spread over much of the earth, culminating in the Ice Age.

The great Ice Age is supposed to have occupied roughly the last 1 million years of geological time. It ended, according to the latest radio-carbon dating, about twelve thousand years before the present, and the period is characterized by repeated fluctuations of mean temperatures of the air and sea water through a range of several degrees centigrade. Four main glacial advances occurred when widespread continental ice sheets repeatedly covered large areas in the Northern Hemisphere, and glaciers were numerous and extensive both in the Northern and Southern hemi-spheres. During three very long inter-glacial periods the climate seems to have been as warm or warmer than at present. Considerable fluctuations of sea-level took place, as well as regional subsidence of the earth's crust under the extra loads imposed by the weight of the ice sheets. Over regions not covered with glacial ice wide fluctuations of rainfall occurred.

Measurements relating to the later part of the Ice Age are mostly confined to reliable carbon—14 datings which extend back to the last 50,000 years. These dates show that the last great glacial invasion of the Great Lakes in North America began more than 25,000 years ago, reaching its maximum extent around 18,000 years ago. Thereafter, the ice sheets shrank, re-expanding at least twice again to reach their maximum sizes 12,000 to 13,000 and 10,000 to 11,000 years ago. The ice sheets then began to shrink. By 5,000 years before the present they may have disappeared completely from the Great Lake Basin in North America. There is evidence to suggest that about this time mean annual temperatures in middle north latitudes were a few degrees above those now prevailing.

Climatic fluctuations in Europe during the Ice Age were similar to those in North America, and much geological evidence has been found in alpine areas pointing to the occurrence of four glaciations separated by three inter-glacial periods. The former have been named Günz, Mindel, Riss and Würm, after certain locations in the Alps where deposits were formed as a result of glaciation. Today, nobody can really assert that the Ice Age has come to an end, or whether we are, in fact, living at the beginning of an inter-glacial period which may be followed about 50,000 years hence by yet a fifth glacial.

What caused the Ice Age is still a matter of speculation, but its effects on the mammal fauna of Europe and North America were profound.

During the early part of the Pleistocene, when the Ice Age had not yet begun, a group of new mammals now known as the Villafranchian fauna appeared in Europe, Asia and Africa. The group included such animals as true horses, the first of the true elephants, and, in parts of Asia, the modern camel. The appearance of these animals is regarded by many as the criterion of the Pleistocene.

About a million years ago, the first of the four principal ice ages began to spread freezing fingers across much of the Northern Hemisphere, forming ice sheets that extended from centres in the north, and from high mountain ranges like the Alps. Between the times of glaciation the extensive warm periods caused the ice sheets to recede to the north again, freeing the lands for new habitation. When the ice sheets formed, taking water from the oceans, sea-levels were generally lower. This exposed new islands and formed land bridges between areas hitherto separated. New migration routes were thereby opened for animals able to withstand the rigour of cold. However, as the ice melted again during the inter-glacial phases, these routes were closed.

During the glacial phases temperatures were generally low over much of the earth, and areas covered by hundreds and thousands of feet of ice could have provided little support for animal and plant life. As the ice sheets advanced, animals retreated southwards to more hospitable areas.

During the First Inter-Glacial, when the retreat of the ice made it possible for mammals to expand their habitats northwards again, there is evidence that mastodons, rhinoceroses, sabre-tooth tigers, elephants and many other large mammals spread as far north as Britain. Some of their remains have been found in the Cromer forest bed in Norfolk, England. They lived there about half a million years ago, but no evidence has been found of them during later geological periods.

As the glaciers waxed and waned under changing climates, and animals had to adapt themselves to temperature extremes, their geographical distribution must have varied to a marked degree. Very probably the temperature fluctuations of the Ice Age speeded up evolutionary processes among animals, giving rise to such mammals as the woolly mammoth, woolly rhinoceros, and certain musk-oxen. Although popularly associated with the glacial ages, these animals in fact did not make their appearance until the end of the Pleistocene in the Last Glacial Phase.

Primitive Man from Java

EUGENE DUBOIS, born in 1858, studied medicine at the University of Amsterdam. In 1886 he received an appointment as lecturer in anatomy. His real interest, however, was vested in the idea that primitive fossil man lay buried in the sediment of the East Indian islands. So convinced was this young Dutch physician that he was right that he resigned his Amsterdam appointment in 1887 and accepted a medical post in the Royal Dutch Army. He set sail for Sumatra in November of that year, and his first excavations brought to light a number of orang teeth and other fossils of fairly recent origin. At about the same time a fossilized skull was unearthed near Wadjak, on the southern coast of Java, being similar in appearance to those of living natives in Australia.

This was not what Dubois was looking for, however. Yet his studies of works on Indonesian palaeontology, and the interesting fossil material which had reached him from Java, convinced him that this latter island held out more promise for continuing his work. Fortunately he was able to obtain a transfer from the Colonial Government and by 1890 his hunch began to pay off.

Examining strata laid down by volcanoes in Central Java about 500,000 years ago, he came upon a small piece of human jaw. The following year he moved his field of operation to the banks of the Solo river where the remains of extinct animals had been discovered. Excavating near the village of Trinil, he found some mammal remains and an upper right molar tooth of human appearance. A month later came the find of his dreams—a partial skull suggesting a brain too big for an ape, yet too small for any known man.

The rainy season halted his explorations for a time, but a few months later he returned to the site to discover, only forty-five feet away but at precisely the same level, a fossil thigh bone (femur), the straightness of which suggested that it came from an erect primate.

Dubois named these fossils *Pithecanthropus erectus*, and described Java Man as a 'missing link' between ape and man. In Greek *pithekos* means ape and *anthropos* man. Its name, therefore, refers to an erect walking ape-man.

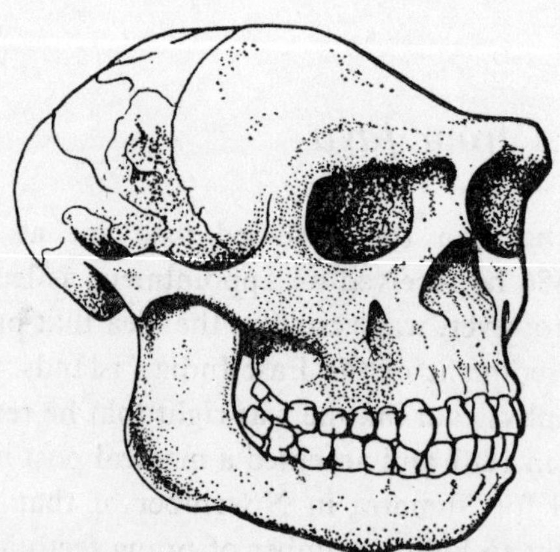

FIG. 72. The skull of Java Man (*Homo erectus*). Reconstructed by Dr F. Weidenreich. (From an illustration in the British Museum (Natural History), by permission of the Trustees.)

Scientific controversy over these fossils became fierce and prolonged. Some anthropologists thought the fossils human, others considered them intermediate between ape and man, while a number pronounced them anthropoid ape. Anti-evolutionists, obviously, had little sympathy for any fossil which might destroy their illusions regarding man's past. For reasons best known to himself but probably connected with the lack of support he had received for his interpretations of the Java fossils, Dubois locked his finds away in strong boxes in the Teyler Museum of Haarlem, Holland, and denied access to everyone for thirty years. It was not until 1923 that he was persuaded to allow fellow scientists to re-examine the fossil bones.

More than forty years passed before further finds of Java Man came to light. First, during geological mapping work near Surabaya in Eastern Java, part of the skull of a human infant was discovered in a new layer of the Middle Pleistocene. It was associated with a new fauna, the Djetis, older than the Trinil fauna where Dubois had discovered his specimens.

Then, a year later, in 1937, the palaeontologist Dr G. H. R. von Koenigswald made important discoveries at Sangiran in Central Java. These included a primitive lower jaw as well as fragments of a skull. All were associated with the Djetis fauna.

Since then, Indonesian scientists have discovered such fossils near Sangiran. In 1962 a jaw was unearthed, followed the next year by the left side and back of a skull, and in 1965 an almost complete skull was bought to light.

Much interest has been aroused by von Koenigswald's discovery of part of a lower jaw from the Djetis zone. It was more massive than any of the previous Java Man discoveries and he named it as belonging to *Meganthropus*. It has now been assigned to a creature probably similar in appearance to a large South African man-ape.

Sufficient fossil material is now available to allow us to form a fairly adequate picture of Java Man's appearance. He must have been about our size, judging from thigh bones which are almost indistinguishable from our own. His thigh bone, and the position of the opening of the skull by which the nerve cord leaves the brain and skull (*foramen magnum*), show that he must have walked upright. He had a strong muscular neck carrying an exceedingly heavy and thick skull. The mouth region was protruding and the jaws, which were on the big side, contained teeth so large that none of a similar size have ever been found in the skull of any kind of man above the level of man-apes. It is, in fact, his dentition that somewhat surprised zoologists, for they found in the upper jaw a toothless space between the canine and incisor like that into which an ape's lower canine fits when its jaws are closed. This so-called diastema is a typical ape-like trait, and Java Man is in fact the only hominid with this kind of space in his dentition. Not even the South African man-apes, generally considered more primitive than *Pithecanthropus*, display this gap. Yet it is certain that Java Man was not closely related to the pongid apes, for his canine teeth, though quite sizable, did not project in a manner requiring such a gap. Probably the opening was simply part of the rather loosely spaced forward tooth row. Not all the Java Man fossils show this gap, which seems largely confined to males.

Compared with our own, his brain was very much on the small side. An endocranial cast made from the first skull Dubois found had a capacity of about 940 cubic centimetres. Later finds were rather smaller than this, compared with ours of 1,450 cubic centimetres, and that of living apes of 500 cubic centimetres. His brain was about 200 cubic centimetres larger than that of the South African man-apes, and thus Java Man is today regarded as the earliest of human kind about whom we have satisfactory knowledge.

Recent potassium-argon datings suggest that the last of these men lived about 710,000 years ago, and since the oldest of his fossil remains date back to about a million years, he must have been around for a considerable time. Much later *Pithecanthropus* appears to have been followed, on Java, by a more advanced hominid with an average brain capacity of 1,100 cubic centimetres. Evidence to this effect was discovered in 1931 and 1932 on the Solo river, at a higher level than the Trinil beds, near the village of Ngandong, where members of a geological survey expedition unearthed many thousands of fossil animal bones, as well as fragments of eleven or twelve human brain cases and two shin bones. This Solo Man (*Homo soloensis*) was clearly of a more recent age and his bigger brain suggested that he may have been more intelligent than his probable ancestral relative, *Pithecanthropus*. With his increased intelligence, however, it seems he developed a number of thoroughly unpleasant habits. His fossilized skulls all have large pieces missing which appear to have been knocked out from below, indicating that this was the simplest way to get at his fellow creature's brains, which he probably enjoyed eating. It is difficult to associate evolutionary progress with such cannibalistic habits, but then we really do not know how Solo Man lived, and what these fossils represent.

Clearly, Solo Man was still under-brained compared with ourselves, but he had a brain about double the size of a gorilla's, and one cannot help wondering whether man's propensity to kill his own kind—so much in evidence and under discussion today—was not in some way associated with the evolution of increased intelligence and the coming of human self-awareness.

Were these perhaps the first signs that the hominids were throwing off the shackles of the random evolutionary experiments of nature?

It is a long way from the Solo cannibals, who might have been able to reason out certain aspects of food value of their own kind, to a carefully planned and executed modern war fought to gain political or economic advantage over another nation. Yet the two situations exhibit certain legitimate parallels which I believe are more than coincidence.

Dr F. Weidenreich (1873–1948), the well-known German anatomist and anthropologist, believed that Java Man was a direct ancestor of Solo Man who lived near the beginning of the Last Glaciation, about a hundred thousand years ago. What happened in the intervening five-hundred thousand or more years, between the time Java Man disappeared from the scene and Solo Man became established, is a matter for pure speculation. The distinct possibility of a direct ancestry between these early men and the fossil sequences found on the island make Java a unique and valuable showcase of hominid evolution.

Primitive Man and his Tools

HORSES HAVE TEETH AND HOOVES suited to a plant-eating animal living on grassy plains; beavers are dependent for their way of life on highly specialized incisor teeth capable of stripping and felling trees; other mammals have specialized in different directions. Man avoided any such specialization and retained the five fingers on his front limbs, so useful to his small tree-dwelling ancestors.

We have previously reviewed Professor Wood Jones's arboreal theory which offers explanations as to the origins of human erect posture and man's ability to use his hands freely to make and manipulate tools. The latter two activities, however, were also dependent on adequate mental powers and bodily co-ordination. A beaver can fell trees by using his powerful incisor teeth as cutting tools, but they are a limited kind of equipment set solidly in his jaws, and cannot be exchanged for improved tools.

With man it is obviously different. Capable of making and shaping his own stone tools, early man soon gained an enormous advantage over his fellow mammals. His limbs and body had remained unspecialized, and therein lies the secret of his special abilities. A part of the story of human cultural development is based on man's successful endeavours to create new tool industries for manipulating his environment, and it is to man's early tool industries that this section is devoted. It will take us beyond Java Man to the men of the Middle and Upper Palaeolithic with whom we shall deal in more detail later in the chapter.

The terms 'culture' and 'industry' are frequently used in connection with Stone Age peoples. Like cultures of modern societies, those of prehistoric man included the sum total of their habits, beliefs and techniques.

Only the durable products of their ways of life have survived through the ages. Of these, tools—particularly those made of stone—are the most numerous examples. A set of artifacts made by a single group of prehistoric man is now generally referred to as an industry.

Unworked bones and stones were probably used as tools by *Australopithecus*, although this has not definitely been proved. Dr Dart was convinced that antelope jaws lined with teeth, and heavy leg bones of animals found in association with the remains of *Australopithecus*, did not get to their living quarters simply through accident, but were taken there by the man-apes and used by them as scrapers. The heavy leg bones were employed as clubs for slaying their prey. Bones and wood unfortunately disintegrate with age and those pieces left behind as fossils may not tell us the whole story of the extent of their use. It is different with stone implements, most of which remain virtually unchanged through the ages. The degree of workmanship and the strata where they are found are often indicative of the cultural development of the people who made them. Prehistoric man learned by experience which stones were best for working into tools. Where possible he chose the hard silicas such as flint, chert, obsidian and chalcedony. In Europe, however, flint was the most common material used, probably because it was easy to find and yielded to the skills of primitive man. It occurs either as so-called tabular flint in chalk, where it forms layers a few inches thick, or it is found in chalk as rows of scattered nodules. Flint actually consists of very fine crystals of silica tightly packed together, the structure being termed cryptocrystalline quartz.

Prehistoric man worked these flint nodules in a number of ways. Either the nodule itself was roughly shaped and used as an implement, flakes being struck from it until the core had the desired shape, or, more often, the flakes themselves were further shaped into tools.

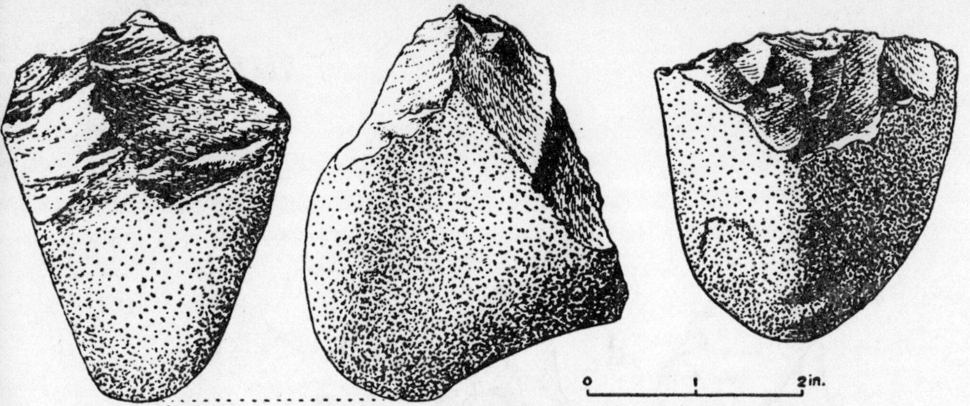

FIG. 73. Primitive pebble tools of the kind associated with *Australopithecus* at Olduvai, Tanzania. (From illustrations in the British Museum (Natural History), by permission of the Trustees.)

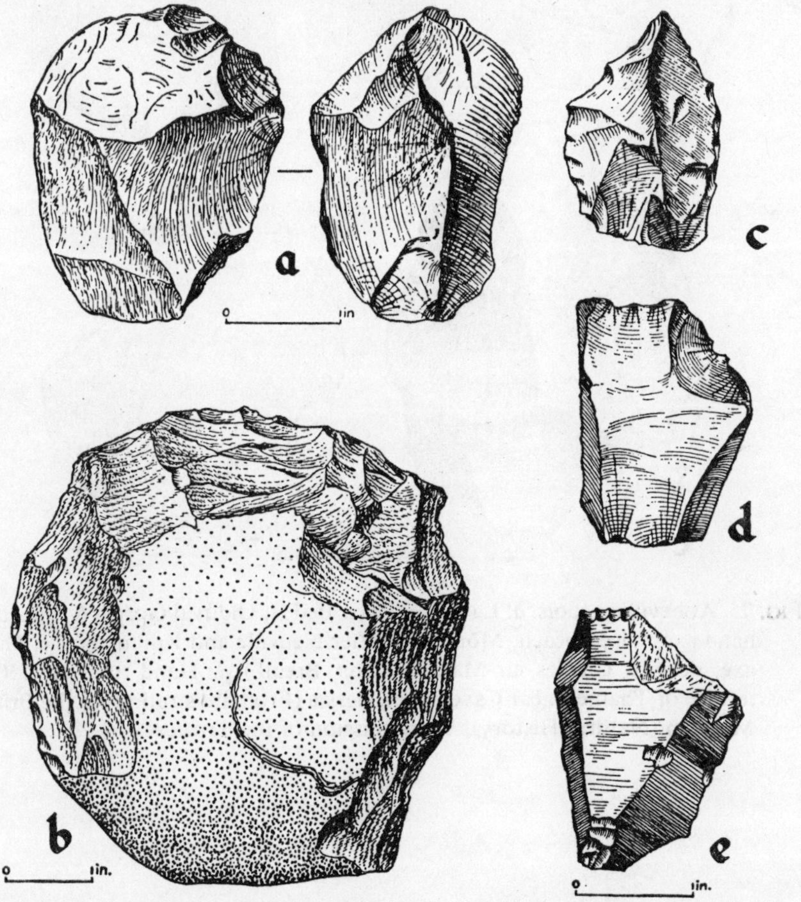

a
c
d
b
e

FIG. 74. Stone tools of Peking Man. *a*. Quartz chopper tool. *b*. Boulder of greenstone flaked into chopper form. *c*. Pointed flake of quartz. *d*. Bipolar flake of quartz. *e*. Bipyramidal crystal of quartz utilized as a tool. (After Pei and Black.)

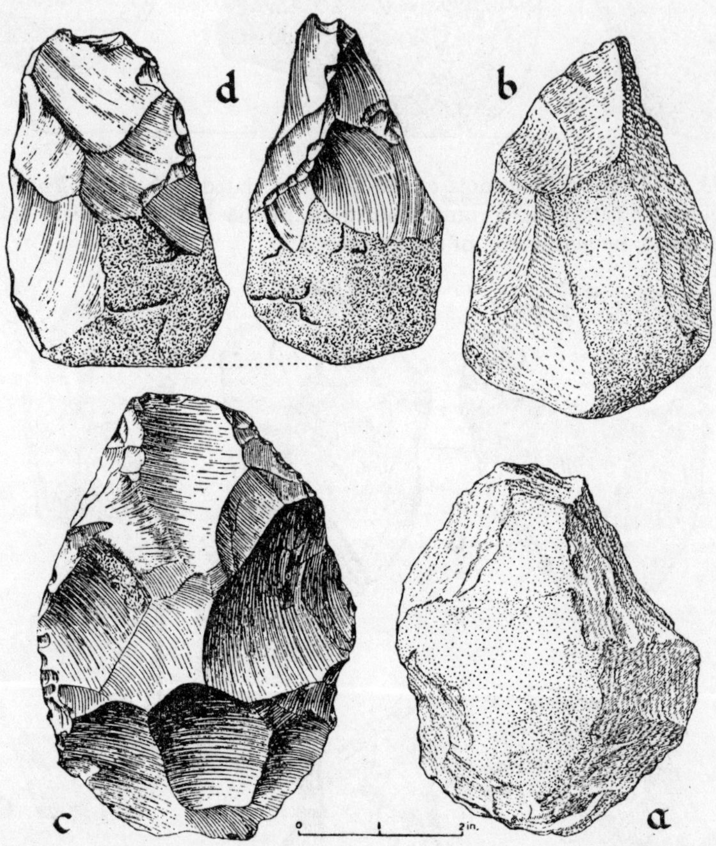

Fig. 75. Abbevillian tools. *a*. Lava hand axe, Bed II, Olduvai Gorge. *b*. Quartzite hand axe, raised beach, Morocco. (After Neuville and Ruhlmann.) *c*. Hand axe, derived Chelles-sur-Marne. (After Breuil.) *d*. Hand axe, on 150-ft. terrace of Thames, near Caversham, Berks. (From illustrations in the British Museum (Natural History), by permission of the Trustees.)

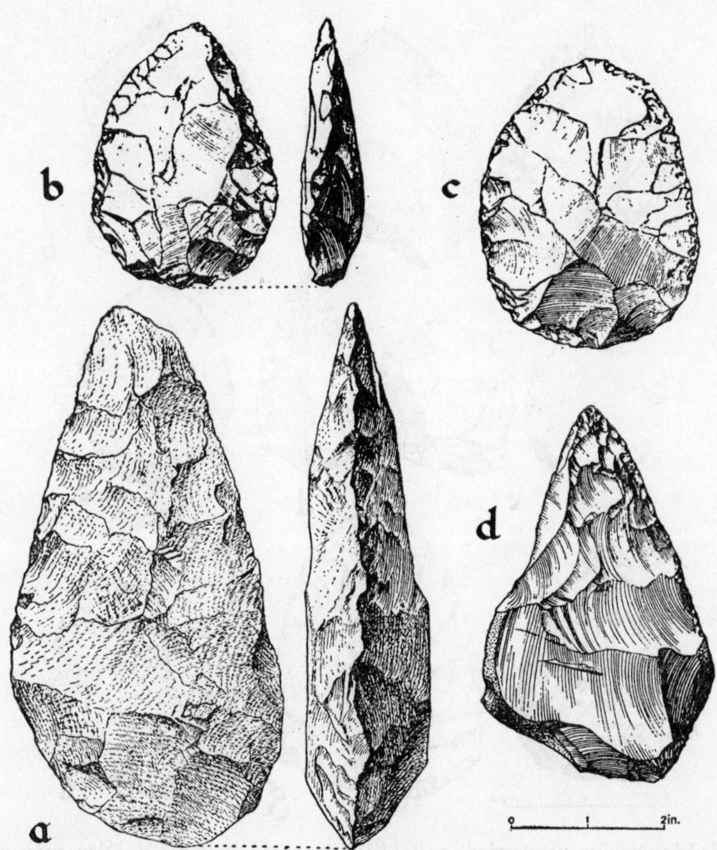

FIG. 76. Acheulian tools. *a*. Lava hand axe, Ol Orgesailie, Kenya. *b*. Twisted ovate, *argile rouge* on 30-metre terrace, St. Acheul, near Amiens (Somme). *c*. Ovate hand axe of flint, South of Wady Sidr, Palestine. *d*. Hand axe, of Micoquian type, below brickearth, Hoxne, Suffolk.

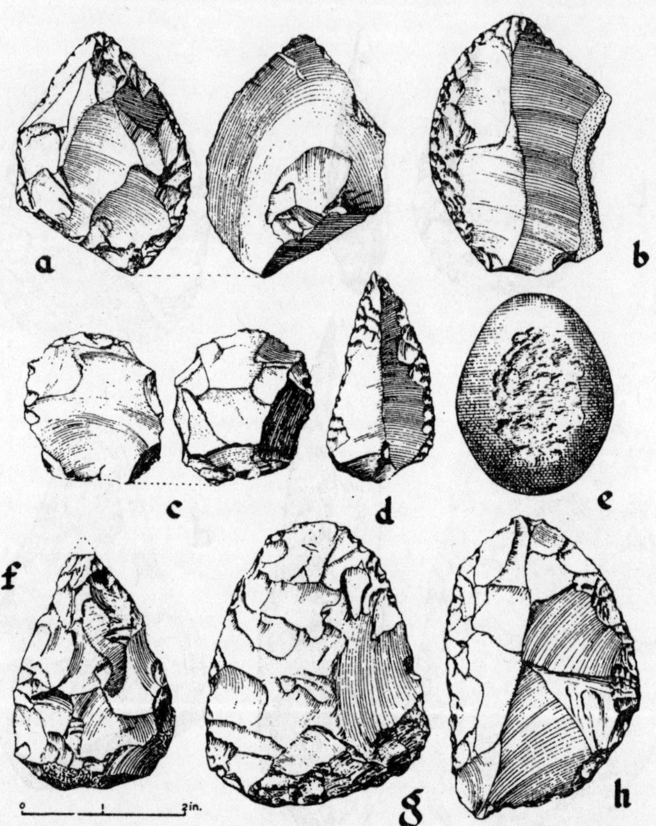

FIG. 77. Mousterian industries. *a*, *b*. Side-scrapers (*racloirs*). *c*. Disc-core, and *d*. point, from rock shelter at Le Moustier near Peyzac (Dordogne). *e*. Small anvil- or hammerstone (pebble of ferruginous grit), Gibraltar caves. *f*. Hand axe from Le Moustier. *g*. Hand axe (chert), and *h*. oval flake tool (flint), from Kent's Cavern, Torquay. *a-d*. Typical Mousterian; *f*, *g*, *h*. Mousterian of Acheulian tradition.

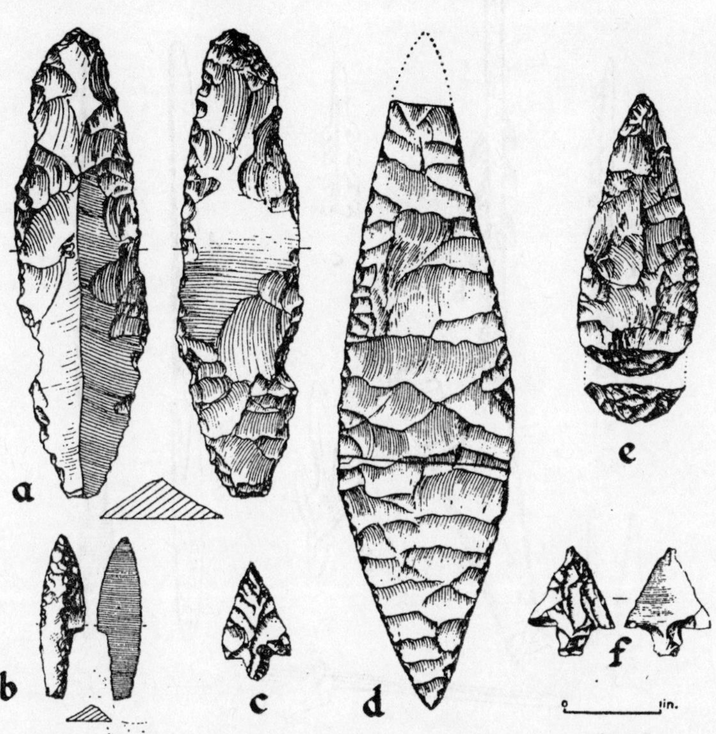

FIG. 78. Late Palaeolithic stone weapon-heads. *a.* 'Proto-Solutrean' point, Ffynnon Bueno, Vale of Clwyd. *b.* Shouldered 'willow-leaf' point (*pointe à cran*), showing pressure-flaking, Solutrean, Fourneau du Diable, Bourdeilles (Dordogne). *c.* Solutrean arrowhead, Parpalló, Spain. (After Pericot.) *d.* Solutrean 'laurel-leaf' blade, or bifacial foliate, Solutré, France. (After de Mortillet.) *e.* Stillbay point (silcrete), made from pointed flake with faceted butt, South Africa. *f.* Aterian arrowhead, Morocco. (After Caton-Thompson.) (From illustrations in the British Museum (Natural History), by permission of the Trustees.)

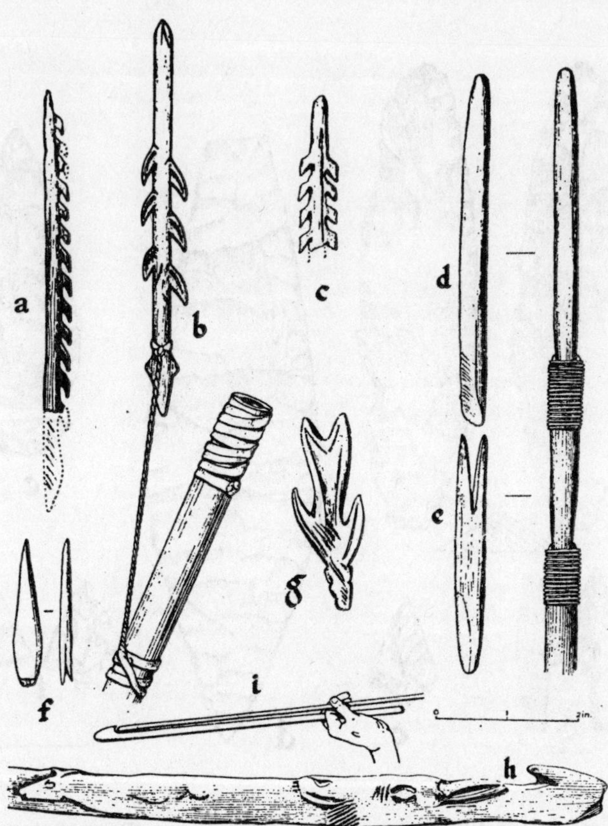

FIG. 79. Upper Palaeolithic bone and antler weapons, *a* to *h*. *i*. Wooden spear-thrower (Womera) in use by Australian aborigine. (From illustrations in the British Museum (Natural History), by permission of the Trustees.)

The Stone Age of man known as the Palaeolithic may have extended from nearly 2 million years into the past to a time as recent as fifteen thousand years ago. Palaeontologists have divided this immense time span into three very unequal parts—the Lower Palaeolithic (the longest), the Middle Palaeolithic and the Upper Palaeolithic. Each of these represents stages in the cultural evolution of man. They fall into the geological epoch of the Pleistocene which is now believed to have lasted nearly 3 million years, ending about ten thousand years ago. We know that much of this period was taken up by the Ice Age, and there is little evidence that man made stone tools before the beginning of the First Glaciation, the Günz, about a million to a million and a half years ago.

These earliest tools, known as 'Eoliths' (meaning earliest recognizable implements) were first described in 1891 by J. Prestwich, after some crudely shaped flints had been discovered by amateur archaeologists on the North Downs of Kent.

A good deal of controversy has surrounded these 'tools' and many believe that at least some may have been formed by natural agencies rather than by the hand of man. Consequently, our knowledge as to man's tool-making ability in Pre-Palaeolithic days is still rather incomplete.

The picture becomes clearer as we enter the Lower Palaeolithic, when Stone Age industries begin to represent definite types of artifacts made during various cultural stages of man. Some of the most ancient implements of this period take the form of primitive flake tools, hand axes, pebble tools, and cores discovered in ancient European river gravels. Over a hundred years ago, important discoveries of this kind were made in terraces along the Somme river and at Abbeville, St-Acheul, and other localities in the Amiens district of France. The Palaeolithic cultures have in most parts of the world been named and classified after stone industries found in North Western Europe, notably in France. Of these industries, the Abbevillian appears to be the most primitive, when man was as yet only able to prepare rather crude hand axes. In the following and closely associated Acheulian culture man's hand axe tradition had markedly improved. It differed principally from the Abbevillian industry by more carefully worked and pointed almond-shaped hand axes which in profile have fairly straight edges.

It appears that during the enormous time span of the Abbevillian and Acheulian cultural stages the hand axe was the most important standardized tool, although there is much evidence in Europe that the Acheulians made implements from waste-flakes, as well as cores. Similar cultural stages to those found in Europe have also been identified in Asia and Africa, most of them being named after the original European discoveries. (No hand axes have been found farther east than India.) Of the African industries the oldest and most primitive deserves special mention having been discovered in the Olduvai Gorge in Tanzania. Here prehistoric man flaked boulders in two directions to form simple cutting, chopping and scraping tools. Such primitive artifacts have also been discovered in lake beds at Kanam in Kenya dating back to the First Glacial Phase, which in that part of Africa manifested itself by periods of heavy rainfall. These simple pebble cultures are very old indeed and some authorities believe them to have originated more than a million years ago in the Pre-Palaeolithic, when *Australopithecus* was still much in evidence.

The Palaeolithic, as we have seen, extended over a period of more than a million years, during which time there occurred many important evolutionary changes in man. In fact it is almost certain that a million years ago a man-ape like *Australopithecus* had already taken the evolutionary road which finally led to *Homo sapiens*, the modern species of man which dominates the world stage today. Clearly during this enormous time span man's cultural advances by no means matched his physical evolution, his progress as a tool-maker remaining strictly limited.

This apparent conservatism in his tool industries probably reflects certain limitations of the human brain, which at this stage had not yet reached a threshold enabling man to use foresight to any degree.

If we pause for a moment and review the time scale of the Palaeolithic it becomes fairly clear that the first six hundred thousand years of the Stone Age saw only limited progress towards more sophisticated stone tools. In modern times, on the other hand, techniques in science and technology rapidly succeed one another, and can be outdated within months of their introduction, to be succeeded by quicker and more efficient techniques. Clearly this state of affairs did not exist in Palaeolithic times, when Abbevillian methods of tool-making hardly changed at all during the six hundred thousand years they were practised. The Acheulian which followed lasted about one hundred thousand years and although man made greater advances in a shorter period of time, matters were still proceeding at a snail's pace compared with today's state of affairs. Their gradual improvement in appearance and usefulness is indicative of an increasing degree of skill used in their preparation but, considering the lengths of time involved, man's early tool cultures developed remarkably slowly.

The Middle Palaeolithic is associated with the Mousterian culture, which derives its name from the rock shelter Le Moustier in Southern France. It is the culture of Neanderthal Man who entered the European scene about a hundred thousand years ago. These men no longer appeared to have made the crude hand axes associated with earlier periods of prehistory. Instead, they were able carefully to shape a variety of flake tools, obtaining these flakes by splitting them off flint cores. These stone tools are of particular interest because they were made by men who had brains as large and larger than our own, although they were probably not as well developed. It is also the last of the stone industries before modern man appeared. Stone cultures do not normally show sharp divisions, more often gradually passing from one to the next, but with the entrance of modern men there are indications that their industries in Europe rather abruptly succeeded the Mousterian. At this stage man had passed a threshold in brain development when further evolution of this organ did not appear to bestow additional survival advantage upon the possessor. Man had become the thinking being he is today. As might be expected the flint work of these Upper Palaeolithic people was of a high standard, enabling them to produce a variety of slim blade tools, many of which were sharp enough along one edge to serve as knives. These were prepared by striking flakes from a previously shaped core and further working these flakes by a method known as pressure flaking. During the Upper Palaeolithic a number of distinct industries can be identified. The Aurignacian culture principally associated with the Cro-Magnons (we shall meet them later) appears to have spread westwards into France, in some places replacing the earlier Chatelperronian culture which probably originated in South Western Asia to spread westwards before the end of the Mousterian. Other cultures, notably the Solutrean and Magdalenian, are recognizable by their particular tool industries. Thus the technique of pressure flaking is much in evidence in the Solutrean, when men produced large finely worked so-called laurel-leaf and willow-leaf blades.

The men of the Upper Palaeolithic were still hunters and food gatherers, but they possessed many more new ideas than their predecessors, enabling them to develop a sophisticated bone tool industry previously little known. They manufactured bone pins or awls and their spears were tipped with sharpened pieces of bone. During later Magdalenian times bone artifacts included such complex weapons as the antler harpoon of which an example was discovered at Kent's Cavern, Torquay, in England.

Clearly in Upper Palaeolithic times man set out on a course of cultural advance from which he has not deviated since. Not all modern men made the same cultural progress but the difference in the rate of cultural advance is largely due to their environment and not to biological constitution. The controversial problem of the inherent abilities of the different races of man is discussed in a later chapter.

For a long time it was generally believed that the people of Abbevillian and Acheulian cultures were entirely dependent for shelter on naturally formed rock shelters and caves.

That men from the Lower Palaeolithic were capable of constructing temporary shelters has been made clear by discoveries made in 1965 at the eastern section of Nice on the French Riviera. There at a site known as Terra Amata ('beloved land') archaeologists unearthed a number of Lower Palaeolithic camp sites strewn with late Abbevillian and early Acheulian stone tools, some of which appear to be at least 300,000 years old.

The shelters themselves were of course no longer intact, but their outline could be traced by the imprint of a series of stakes averaging about three inches in diameter that must have been driven into the sand to form the walls of the house. There was also evidence of a line of stones parallel to the stake imprints which men had placed there to brace the walls. The huts were all of the same oval elongated shape, ranging in length from 26 to 29 feet with a width of about 20 feet.

Evidence from the camp site indicates that the hunters who visited these Mediterranean shores some three hundred thousand years ago returned there annually but stayed for only relatively brief periods of time.

During their stay they hunted the surrounding countryside and, as is evidenced by fossils, they showed a preference for big game which included extinct forms of stag, elephant, wild boar, rhinoceros and wild oxen. The tools were manufactured on the spot and the tool-makers' 'workshop' was easily identified by stone fragments that surrounded a bare patch of floor where the tool-maker must have sat.

There were also fire pits protected from the prevailing north-westerly wind by a wind screen built of rocks and pebbles.

The site also bears evidence of a number of bone tools; one leg bone of an elephant was found having been hammered at one end. Yet another bone had a fire-hardened point, a technique still used today by peoples of primitive cultures who harden the tips of wooden spears. Other bone artifacts also appear to have been used as scrapers.

Three hundred thousand years ago the climate on the Riviera was temperate, though a little colder and more humid than today, and we might wonder whether these summer visitors found the Mediterranean shores of the Middle Pleistocene a welcome change from the harsher conditions that prevailed during the latter part of the Mindel Glaciation in more central and northern parts of Europe. How far these men travelled we cannot tell, but there is some evidence that not all tools were made locally, some having been brought from more distant places.

This Lower Palaeolithic camp site at Terra Amata bears witness in a vivid and realistic way to the manner in which primitive man lived, hunted and worked, and it is suggestive of an advanced social organization. The French historian Camille Jullian wrote soon after the Terra Amata living floors had been exposed:

> The hearth is a place for gathering together round a fire that warms, that sheds light and gives comfort. The tool-maker's seat is where one man carefully pursues a work that is useful to many. The men here may well be nomadic hunters, but before the chase begins they need periods of preparation and afterwards long moments of repose beside the hearth. The family, the tribe will arise from these customs, and I ask myself if they had not already been born.

The Discovery of Peking Man

DR J. G. ANDERSON, a Swedish geologist, had worked for the Chinese Government as a mining adviser since 1914. For some time he had been interested in the fossil wealth of China and his attention was drawn to this region by a human tooth displayed in a Peking drug store.

In 1921, together with his Austrian associate Dr O. Zdansky, he examined a limestone-filled cave in a hill near the village of Choukoutien, some thirty-seven miles south-west of Peking. There a surprise awaited them for they found several small pieces of quartz that obviously did not belong to the local strata. Since animals are unlikely to have taken the quartz there and man must have found this hard material useful for making tools, the obvious conclusion was that at one time primitive man had lived in this locality. Accordingly they set about looking for him, and after two years of searching in the stratified deposits which had yielded a mass of animal fossils, Zdansky discovered two human teeth. This was scant reward for their labours, but it proved sufficient incentive for palaeontologists to go on looking for more, and more they found as they excavated deeper into the cave. At this stage Dr D. Black, professor of anatomy at the medical school in Peking, became interested, and after examining a third such tooth, found in 1927, he concluded that it was of primitive human origins. Dr Black christened this particular man, of whom only three teeth had so far been found, *Sinanthropus pekinensis* (Chinese man of Peking).

When in 1929 the Chinese palaeontologist Dr Pei was in charge of the excavations at Choukoutien, it was not long before he discovered the first skull still partly embedded in a hard rock covered by loose sand. On the following day the skull, together with the matrix in which it was embedded, was carefully excavated and later taken to Peking. Clearly the fossil was prehistoric and immensely valuable to anthropology. The press received Peking Man rather more friendlily than his Java relation, and fellow scientists quickly acknowledged his importance. The first cave and other sites at the same hill continued to yield interesting fossil fragments of Peking Man together with his tools. Tragically in 1934 Dr Black, who had been the guiding light in the important Choukoutien excavations, died of a heart attack. Later, in 1935, the Rockefeller Foundation, which was supporting the work financially, invited Dr Franz Weidenreich to carry on Black's work. Under his direction portions of seven skulls were found, and by the time the work was at an end parts of well over three dozen different individuals had been taken from the excavation sites. None of the skulls discovered was complete but by piecing together the various specimens a most the entire cranium could be reconstructed.

Although Peking Man had lain for more than half a million years safely buried in his limestone cave he met with dire misfortune soon after his discovery, for he became one of the first casualties of the war in the Pacific. His remains were lost in the confusion and nobody really knows what has happened to them since. Fortunately in 1941 Dr Weidenreich took a set of beautifully made Chinese plaster casts back with him to New York where they remain as the only evidence of the lost fossil treasure.

Peking Man forms an important link in man's ancestry. He is closely related to Java Man and virtually identical in some respects, but he was a brainier creature and his teeth were more like our own. It is noteworthy that the skeletons both of Peking Man and Java Man, as far as they are known, can hardly be distinguished from ours, and they exhibit none of the differences in form seen in the African man-apes (australopithecines). Their skulls, however, were rather heavy and thick as compared with ours, with large continuous supraorbital ridges (bony brows) and a low retreating forehead which, however, in Peking Man showed a little more prominently in the front.

Larger than Java Man's, Peking Man's brain, according to the endocranial capacities of the various skulls, ranged from about 900 cubic centimetres to 1,200 cubic centimetres, suggesting an average figure of 1,050 cubic centimetres. (Our brains average 1,450 cubic centimetres.)

His teeth showed evolutionary progress, too. They are smaller than those in Java specimens, giving his mouth a much more human appearance. Peking Man also lacks the gap in the tooth row to receive the canine from the opposite jaw, found in the male skulls of *Pithecanthropus* from Java. The dentition indicates that his diet consisted both of vegetation and meat. In his time natural selection no longer favoured very powerful jaws, for he made tools to cut his food and, what is more, there is clear evidence from his homesteads that he had mastered the use of fire and therefore probably cooked his food. And this was important for him. Whereas all the higher primates have adapted to foods that are largely vegetable, man's digestive system can also cope with a diet of meat and fat, especially if the meat has been cooked. Fire may therefore have been principally important to early man as a means of supplementing his diet with meat, a food of high calorific value of which considerably less is required to sustain the body's metabolism than if only vegetables are eaten. The modern gorilla has to spend most of its time feeding to sustain its bulky body on an entirely vegetarian diet. Thus meat-eating probably greatly added to the general efficiency of prehistoric man.

Still a relatively under-brained creature, Peking Man was, nevertheless, an accomplished tool-maker and -user. His cave homesteads yielded thousands of chipped-stone tools of less primitive construction than the Oldowan pebble tools found in Bed I of Olduvai Gorge in Tanzania. They clearly belonged to the Abbevillian industry. He had used quartz and greenstone to make chopper artifacts and there were also pointed flakes of quartz struck from a core and edged scrapers that might have been useful to clean the skins of animals. Whilst there is evidence that he used some animal bones as tools he did not appear to have adapted them specially before use.

Probably one of the reasons why so many of his skulls have become available for study was that he, like Solo Man, was a cannibal who enjoyed eating the brains of his fellows, leaving many parts of their skulls littering his homestead.

This rather unfortunate tendency of early man should, I believe, not be over-emphasized, because apart from his cannibalistic habits, which doubtless he possessed, he must also have been endowed with a social organization considerably in advance of that of any of our living apes. This allowed him to exercise powerful influence over his own kind and the creatures which shared his environment. If Solo Man and Peking Man had spent most of their time roaming through Pleistocene forests murdering and eating their own kind, natural selection would soon have disposed of their species, and they would not have survived the 300,000 years or more during which they inhabited large parts of the Old World.

In 1963 Chinese anthropologists made a number of further important discoveries. Near Chenchiawo in Shensi province they discovered a primitive human lower jaw in good condition, while a year later excavations on Kungwangling Hill about thirteen miles away yielded a well-preserved skull cap, and parts of a face and a tooth, all of which appeared to have belonged to the same skull. These specimens are believed to be older than *Sinanthropus* and the owner may have been a contemporary of the oldest Java Man. These finds were particularly important because they represent a near relative of Java Man who lived well up on the Asian mainland. They also indicate a small degree of evolutionary progress between the earlier and later of the Asian fossil men. Thus, Java Man and his Chinese mainland contemporary are the earliest and most primitive known types of man, followed by Peking Man, and long afterwards Solo Man on Java, who appears to have been a much later relative of the same group, having advanced only relatively little during his much longer evolution. These explanations may be over-simplified, for they are based on a very scant fossil record, yet it seems fairly clear that only one general kind of man occupied the whole Far East during the Middle Pleistocene.

Early Man in Europe and Africa

ASIA AND THE INDONESIAN ISLANDS yielded a fine collection of skulls dating back to the Middle Pleistocene, whereas the tools left behind by Java Man were insufficient to enable us to find them, though there is little doubt that he was capable of making and using them.

On the other hand palaeontologists were able to trace a longer story of human tools in Europe and Africa, although before 1950 they were faced with the frustrating lack of early human remains—with one major exception. This was the discovery in 1907 of the Heidelberg jaw—a find that can be ascribed entirely to the great enthusiasm of Dr O. Schoetensack of Heidelberg University, whose search for early human remains had stretched over the previous twenty years. The discovery was made near the little village of Mauer a few miles from Heidelberg, where the river Neckar had laid down huge sand deposits that were worked for commercial purposes and had yielded numerous fossils, including bones of ancient elephants, lions and rhinos. Among these animal fossils a large human jaw was found at a point about eighty feet below the surface.

Heidelberg Man may have been a contemporary of the earliest Java Man, while the last of the man apes of Africa were still flourishing. He was, therefore, a very ancient member of the human family who probably lived at the beginning of the Second Glaciation (Mindel) some 800,000 years ago.

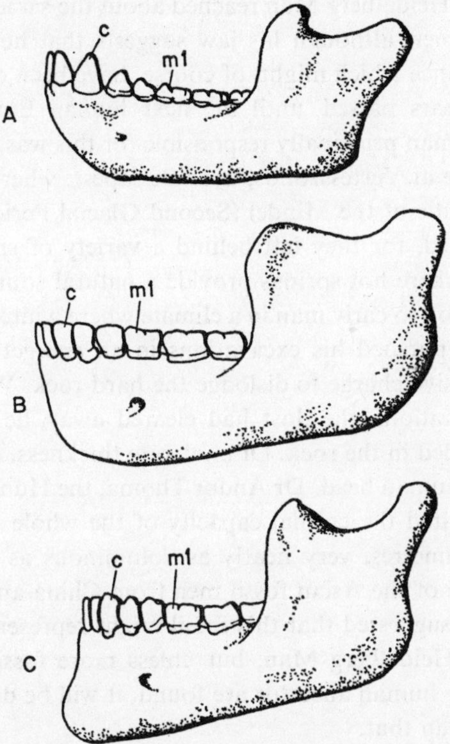

FIG. 80. The Heidelberg mandible (B), compared with the mandible of a chimpanzee (A) and modern man (C). In each case the canine tooth (c) and the first molar tooth (m1) are indicated. (From illustrations in the British Museum (Natural History), by permission of the Trustees.)

The most striking feature of this fossil jaw is its huge size and heavy build. It has a somewhat ape-like appearance because it lacks a chin prominence, but the dental arch and the teeth point distinctly in a human direction and the latter are decidedly small and nearer to modern man than they are to Java Man. The canines, well reduced in size, do not project at all above the other teeth, and the molars too are of more modern appearance, showing wear indicative of the chewing motion of humans rather than of apes.

Heidelberg Man's skull remains somewhat of a mystery, for apart from this one lower jaw no other portions of his skeleton have ever been discovered, and this one item provides insufficient guidance to attempt a reconstruction of his head. A probable contemporary of Java Man, it can be assumed that Heidelberg Man reached about the same stage of development as the former, although his jaw suggests that he had a somewhat different appearance which might, of course, have been of a regional kind.

Fifty-eight years passed until the next human European fossil was discovered. The man principally responsible for this was Dr Laszlo Vertes, who found a site at Vertesszöllös, near Budapest, where during the mild break in the middle of the Mindel (Second Glacial Period) primitive men must have camped, for they left behind a variety of small pebble tools. This is an area where hot springs provide a natural source of warmth, no doubt an attraction to early man in a climate where winters must have been cold. Dr Vertes pursued his excavations in an energetic way, using the occasional explosive charge to dislodge the hard rock. When, after one of the blasting operations, the dust had cleared away, he observed a bone fragment embedded in the rock. Of moderate thickness, it was clearly part of the back of a human head. Dr Andor Thoma, the Hungarian palaeontologist, has estimated the cranial capacity of the whole skull to be about 1,400 cubic centimetres, very nearly as voluminous as our own and far larger than those of the Asian fossil men from China and Java. Professor W. Howells has suggested that this fossil might represent the back of the head of a later Heidelberg Man, but unless more fossil remains of this highly interesting human ancestor are found, it will be difficult to take the matter further than that.

Let us now travel to North Africa to investigate a site near Ternifine in the Oran region of Algeria. Here, at the onset of the Mindel, about 800,000 years ago, the sea-level was beginning to fall, resulting in the formation of a pond. During later eras, the whole depression silted up to become completely filled with beds of sand. Today, it has become a commercial sandpit. Here, in 1931, archaeologists discovered primitive stone implements along with the bones of antelopes, giraffes, zebras, rhinos and elephants, as well as carnivores of various kinds. Working under the constant threat of flooding, Professor Camille Arambourg began systematic excavations there in 1954 and 1955. His labours were eventually rewarded by the discovery of three human jaws and a parietal bone (the bone that forms the side of the cranium). Some difference of opinion exists among anthropologists whether these fossils belong to a new kind of man and should therefore be accorded a separate genus, but the majority opinion tends to the view that his features were close enough to Java Man and Peking Man to be classified with them.

Soon afterwards, further excavation work in Barbary brought to light a fragment of another and almost identical jaw near Casablanca, Morocco, originating from a somewhat later period.

A further early man belonging to the same period has been discovered in Africa. It was found by Dr Leakey, who, in 1960, came upon a complete skull cap in Bed II of Olduvai Gorge. With it were stone artifacts similar to those discovered at Ternifine. The skull has been judged to have contained a brain of about 1,000 cubic centimetres, similar to that of Peking Man. This discovery provides the most positive confirmation of the existence of men like Java and Peking Man in Africa and it is particularly significant that this find was associated with a tool-making industry.

A Brief Summary

Homo erectus is a name applied today to all those early men who lived from about one million to half a million years ago, including Java Man, Peking Man, and the European and African fossil men that fall within that period, but excluding Solo Man.

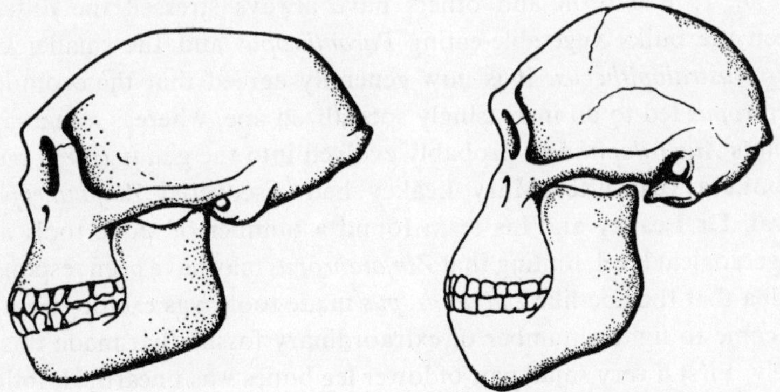

FIG. 81. Skulls of *Homo erectus* (left) and *Australopithecus* (right). (After Bernard Campbell.)

The evaluation of these various fossils has possibly created an artificial and over-simplified picture. Nevertheless, the evidence shows that men of similar stature and appearance lived in many parts of the Old World during some half a million years. It justifies the inclusion of all of them in the single species *Homo erectus.*

Another New Species?

THE PICTURE that has so far emerged reveals three reasonably distinct landmarks in the evolution of our species. First we have *Ramapithecus*, a possible first hominid, then follow the australopithecines which include *Australopithecus* and *Paranthropus*, of which so much evidence has been found in Africa. The former were probably omnivorous and capable of using tools whereas the genus *Paranthropus*, which included a number of gorilla-like man-apes, were vegetarians and unconnected with any tool culture. Mary Leakey's discovery of the *Zinjanthropus* skull at Olduvai demonstrates that the australopithecines were widespread 1,750,000 years ago, and they certainly left useful evidence of their whereabouts in limestone-filled caves right up to about a million years ago. Anthropologists are now trying to solve the riddle of how *Homo erectus* evolved from a man-ape that could have been an improvement in the human direction on the *Australopithecus* we know from fossils. However this may have happened, it must have been a slow process, like most forms of evolution extending over 1 or 2 million years.

Dr J. T. Robinson and others have always stressed the difference between the bulky vegetable-eating *Paranthropus* and the smaller omnivorous *Australopithecus.* It is now generally agreed that the evolution of *Paranthropus* led to an increasingly specialized ape, whereas a man-ape of the genus *Australopithecus* probably evolved into the genus *Homo erectus.*

About a year after Mary Leakey had discovered *Zinjanthropus* at Olduvai, Dr Leakey and his team found a number of stone tools at the same geological level, hinting that *Zinjanthropus* may have been responsible. The idea that the ape-like *Zinjanthropus* made tools was exciting; but then there came to light a number of extraordinary fossils that made this idea unlikely. First a very small pair of lower leg bones was unearthed, followed by the discovery a little lower down in Bed I of a lower jaw, a number of bone fragments belonging to a hand and part of two parietal bones belonging to a much smaller and very different primate than *Zinjanthropus*. Finally, fossil parts of at least six such individuals were identified, originating from various levels at the Olduvai Gorge excavation site, so that it was possible to determine their relative positions to one another and to *Zinjanthropus*.

Evaluation of these bones established that their owners were completely bipedal. The skull revealed a larger and rounder brain case than that of *Australopithecus*, with lighter jaws and teeth markedly smaller than those of the man-apes. It was safe to say that these fossil bones must have belonged to a creature decidedly more progressive than *Australopithecus* and *Zinjanthropus* (*Paranthropus*) but also very different from *Homo erectus*, with his thick, low-angled cranium.

Dr Leakey and his associates were convinced that they had discovered a new species of primitive man and accordingly they named him *Homo habilis*. Similar to *Australopithecus* in build but with a better developed brain, he might well have been responsible for the stone implements found in the same bed at Olduvai; and this, according to some authorities, would qualify him for inclusion within the genus *Homo*. As is often the case with fossil finds, anthropologists placed different interpretations on them. But later this controversy was partially resolved by further discoveries suggesting that this now famous Olduvai site harboured numerous stages of hominid evolution running from a creature akin to *Australopithecus*, the South African man-ape, to more advanced forms of men now included in the species *Homo erectus*. As matters stand at present, it seems that Peking Man and his relatives now all included in the species *Homo erectus* had been widespread in the Old World during the Middle Pleistocene and that *Australopithecus* and his relatives preceded them during earlier ages. The time lapse between the two is roughly a million years, and since no evidence exists during this interval of other completely different hominids, we have to rely upon available fossils to bridge the gap between these two very different species.

Professor W. Howells supports the view that *Australopithecus*, a slightly built omnivorous feeder and hunter, probably capable of using the bones of animals for tools, gradually evolved into the early men assembled in the species *Homo erectus*. Successful evolutionary experiments first increased brain size, the body size remaining small—*Australopithecus* was only about four feet tall. Later natural selection also favoured an increase in body size, probably for reasons of more effective defence similar to those which operated in the evolution of the gorilla-like *Paranthropus*. These changes were accompanied by alterations in skull form and thickness. Many anthropologists doubt the necessity for the inclusion of the new species, *Homo habilis*, into man's line of evolution, and Dr Robinson has even suggested that *Australopithecus* should be allowed into the genus *Homo*.

Vegetarian gorilla-like creatures now classed in the genus *Paranthropus* appear at one time to have been widespread, but were probably never in evolutionary competition with *Australopithecus*. There is no evidence that they ever used tools and they appear to have died out in the Middle Pleistocene.

The Neanderthal Men

We now find ourselves near the end of the Second Glacial Phase, about 500,000 years ago, when the various species of *Homo erectus* still roamed over much of Asia, Europe and Africa. The Great Inter-Glacial Phase began at about this time, followed some 200,000 years later by the Third, or Riss, Glaciation. During all this time there is little evidence from bone remains of man's further progress. Then, about 125,000 years ago, during the warm times of the Last Inter-Glacial, human fossil evidence dramatically revealed a widespread human type in Southern Europe and Western Asia now known as Neanderthal Man. They occupied the human stage about 100,000 years, until the mild phase of the Last Glaciation (Würm). Then, almost miraculously, they disappeared from the scene, to be replaced by several kinds of men of the modern type, *Homo sapiens*. These changes were spread over many hundreds of years, but seen against Neanderthal's history of a hundred thousand years or more, their sudden disappearance suggests that certain catastrophic events may have befallen them. Could this catastrophe in part, at least, have been an early manifestation of modern man's control over his environment?

The real truth about the Neanderthal's disappearance may never be known, but it is reasonable to speculate that wherever modern man set foot he successfully competed with his hominid rivals, which had not quite reached his stage of development, and eventually displaced them everywhere. How this happened we can only surmise, but it is unlikely that the two types of man did battle to the death. More probably the newcomers replaced their cousins by more successful hunting and an advanced social organization.

Neanderthal fossils were first discovered in Western Europe. Later it became clear that they also occupied large areas in Eurasia and North Africa. Rather short and heavily built, these barrel-chested men must have been exceedingly strong, and like Peking and Java Man they had big jutting brow ridges running across the forehead, a protruding muzzle-like mouth, poorly developed chins and a retreating forehead. Males appear to have reached a height of about five feet four inches and the females were somewhat shorter. The Neanderthal skulls, however, indicate that these men possessed brains equal in size to our own, sometimes even larger. Although his great brain had a poorly developed frontal part, which may not have raised him to our level of intelligence, these men were nevertheless accomplished tool-makers and by the beginning of the Würm Glaciation a definite set of stone artifacts known as the Mousterian industry can be recognized as their handiwork.

Despite the fact that in brain size the Neanderthals are our equals, many aspects of their lives must have been severely limited. There is for example no evidence of personal decoration or art and even the perforated teeth of wild animals so often used by hunting communities for adornment are entirely absent from their homesteads and burial places. It appears that the Neanderthals, like the men that lived before them, occasionally ate their own kind, for in the Croatian Cave of Krapina there were discovered the remains of more than ten Neanderthals that appear to have been partially burned.

Evidence from their burial places indicates that they had already advanced spiritual ideas. In this respect the remarkable burial place found in the Cave of Teshik-Tash in Eastern Uzbekistan is particularly noteworthy. Here a child was buried closely surrounded by a ring of five or six pairs of goats' horns still attached to the skulls of the animals and pushed point downwards into the floor.

Other graves in the floors of caves made during the early part of the Würm Glaciation show definite evidence that complete bodies had been buried there, together with a supply of stone tools and other items.

The long calendar of Neanderthal finds begins with the discovery of a child's skull in a cave at Engis, near Liège, Belgium, in the late 1820s. This was followed by the finding of a female fossil skull in a quarry on the north face of the rock of Gibraltar in 1848. Neither of these two finds, unfortunately, were recognized at the time for what they really were.

The sensation came in August 1856 when workmen uncovered human fossil bones in a limestone cave known as the Feldhofer Grotto in the

Neander Valley, in Germany, through which the small stream Düssel flows. Believing they were bones of animals the workmen discarded them, but fortunately the owner of the cave, a Mr Beckerschoff, recovered at least some for posterity. They were given for investigation to Dr Johann Carl Fuhlrott, a teacher at the local Secondary School in Elberfeld, who as an amateur palaeontologist had been collecting specimens from the district for a long time. The remains included such valuable parts as the bones of arms and thighs, fragments of ribs, a skull top and part of the pelvis.

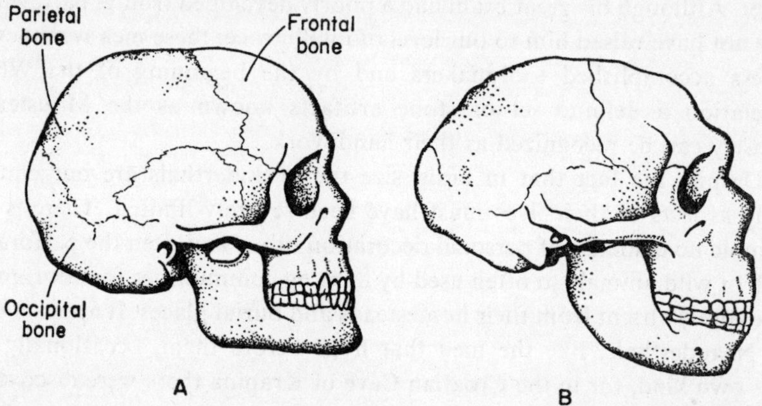

FIG. 82. Skulls of two sub-species of *Homo sapiens*. (A) Cro-Magnon Man. (B) Neanderthal Man.

The controversy that was to surround these fossil bones proved intense, partly because it seemed impossible to associate the finds with stone artifacts or a fauna, so that at the time no definite date for them could be established. Some thought they were simply the remains of a recently deceased man, others made different guesses, mostly in line with the then still current ideas that modern man could never have evolved from what appeared to be an inferior kind of human being.

It is noteworthy that in his book *The Descent of Man* Darwin refers only briefly and cautiously to the Neanderthal remains. Like a number of other students, he kept an open mind on this controversial subject.

In March 1908, Otto Hauser, a Swiss antique dealer, discovered a complete Neanderthal skeleton of a child in a rock shelter at Le Moustier in Southern France. Later the Neanderthal cultural stage was named after this locality—the Mousterian. During the Second World War this skeleton unfortunately was destroyed by a bomb, together with another important find, the Combe Capelle skeleton belonging to a later man, about whom we shall hear more in a moment.

We now journey to the shores of the Mediterranean in Israel, where in 1931 and 1932 an expedition undertook important excavations on the western slopes of Mount Carmel. There in two caves, together with stone implements belonging to the Mousterian industry of European Neanderthal Man, archaeologists discovered fossil parts of more than a dozen individuals, among which were a number of complete skeletons.

In the cave of Mugharet et Tabūn the complete skeleton of a woman was recovered who, despite certain differences—including a more rounded back part of her head, and a generally lighter build—showed marked Neanderthal characteristics. Other skeletons found in the cave of Mugharet es-Skhūl, however, were of a type intermediate between the typical Neanderthal Man and modern man, showing many combinations of the features of both. We will return to our distant cousins on Mount Carmel later to examine some of the explanations that have been given regarding their origins. Other finds in Israel included that made by a Japanese team in 1961 in the Amud Cave, where another good skeleton of a man was discovered.

Earlier, in 1953, while excavating in the Shanidar cave in the Zagros mountains of Northern Iraq, Dr Ralph Solecki of the Smithsonian Institution discovered the remains of a child showing signs that it would have grown up to be a Neanderthal. This discovery was presently followed by six other skeletons of the same kind and more successes came in 1957, when his team discovered three badly damaged adult skeletons in the same cave. Carbon-14 datings have placed these remains at various times from 70,000 to around 50,000 years before the present. Further finds in Belgium, France, Spain, Italy and Russia at last established Neanderthal Man as an early form of man who had occupied large areas of present-day Europe, North Africa and Asia, but not, it seems, England.

Modern Man Enters Europe

THE TIME is now about 35,000 years before the present, a relatively mild spell towards the end of the Fourth Glaciation, when the ice began its final retreat to the north.

The Neanderthals were vanishing into obscurity to be replaced by men of our own sub-species, *Homo sapiens.*

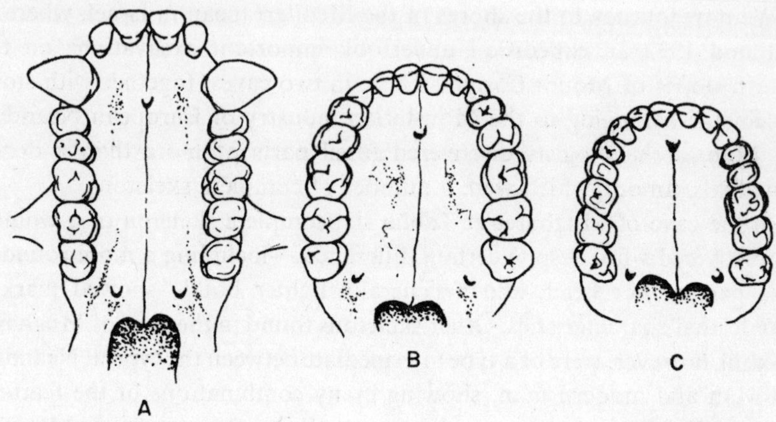

FIG. 83. The palate and upper teeth of (A) a male gorilla, (B) *Australopithecus*, and (C) man. (From illustrations in the British Museum (Natural History), by permission of the Trustees.)

For the next 25,000 years or so, these Upper Palaeolithic men, who appeared to look entirely like ourselves, led the lives of nomadic hunters. In Western and parts of Central Europe they appear to have made their homes in caves and rock shelters, but over many parts of Southern Russia they settled on the banks of large rivers that flowed into the Black Sea, constructing artificial dwellings to protect them against the severe cold of the Last Glacial Phase. In the light of recent discoveries near Nice on the French Riviera, where primitive men had built temporary shelters nearly 300,000 years ago, it is reasonable to suppose that their Upper Palaeolithic cousins who lived more than 200,000 years later must have been capable of more sophisticated construction works, although unfortunately only scant evidence of this exists. Their stone industry succeeded that of the Mousterian culture associated with the Neanderthals, and as earlier mentioned, this Upper Palaeolithic industry which marked the last part of the Old Stone Age showed distinct and more rapid advances in stone tool-making. These men, however, were still food gatherers and subsisted primarily by hunting. After the first period of cold of the Last Glaciation, the glaciers moved back northward, the arctic conditions being replaced in turn by a tundra and steppe fauna and flora. The open country was inhabited by bison, horses and musk-oxen, while the cooler regions were the home of the great mammoth, the woolly rhinoceros and many other animals. Working with his fellows in groups, man had mastered the art of hunting these large and dangerous animals, and possibly the commonest method of capture was to stampede a herd of animals over a cliff or into a narrow ravine. Besides being a source of meat, the animal skins supplied man with clothing and shelter, and the bones and ivory made possible further rapid developments in tool-making. Fish must have been plentiful in the rivers and these Upper Palaeolithic men appear to have developed bone harpoons for catching them.

Their dead, like those of the Neanderthal men, were carefully buried, often decked in the garments worn in life, and sprinkled with red ochre. Fear of the dead does not appear to have been a significant factor, for they often buried their dead by digging graves in the floors of their cave homesteads, afterwards continuing to occupy the same site.

Equally important is the truly remarkable evidence of their artistic ability.

More than a hundred painted or engraved caves have been discovered, usually representing animals and scenes from the hunt. The greatest concentration is in South-Western France and Northern Spain, but isolated caves have been discovered in Central France, Central and Southern Spain, Italy, Sicily, Germany and North and South Africa. Executed in black, red, brown and yellow pigments, many of the paintings are highly decorative and show great liveliness.

Their art, however, appears to have had definite functions, and the animals depicted in the caves may have been intended to influence magically the success of the hunting expeditions upon which the success of the community depended. These artistic achievements show great sensitivity of observation and technical ability.

No reliable method of dating cave art has yet been found but modern methods of radio-carbon dating place some of the earliest examples at between twenty and thirty thousand years old.

Their greater accomplishment as hunters and artists is to some extent reflected in the size and shape of their skulls. They certainly had large heads of modern form and skulls belonging to them have been found with cranial capacities in males of 1,700 to 1,800 cubic centimetres, but there were sexual differences and those of females appear to have been smaller.

Modern man has an average brain size of 1,450 cubic centimetres. Yet to base speculations of early man's intelligence purely on the size of his skull—and this is all we have to go on as far as fossils are concerned—may also be misleading, for nobody is ever likely to know for certain how far his brain had actually developed in his large cranium.

FIG. 84. Upper Palaeolithic cave art. (From illustrations in the British Museum (Natural History), by permission of the Trustees.)

In stature, this race of men was markedly taller than the Neanderthals, many of the males approaching six feet in height. The structure of their skeletons and the shape of their skulls were without any trace of primitive Neanderthal characters.

The first of these new men to be recognized as such was discovered in 1868, when workmen engaged in constructing a railway near the village of Les Eyzies in South-Western France unearthed remains of more than five human skeletons in the rock shelter of Cro-Magnon. Animal bones, sea shells in the form of necklaces, and stone tools were also found. The site was carefully excavated by the geologist Louis Lartet, who published his findings in the same year.

These discoveries aroused much public interest, particularly because some time earlier T. H. Huxley had cautiously dismissed the Neanderthal finds as an extreme variant of the modern type of man—a view shared by most contemporary anatomists.

So much publicity surrounded the Cro-Magnon finds that the work of excavation was carried out under close public scrutiny, supervised by skilled archaeologists. But the Cro-Magnons were not the first of these new men. Their earlier relatives now named after the famous Combe-Capelle skeleton (discovered in August 1909 in the same region of Southern France where Neanderthal fossils had come to light) appear to have been somewhat smaller in body size than the Cro-Magnons, with long, high and rather narrow heads. Similar skulls and skeletons were also recovered near Lautsch in Czechoslovakia.

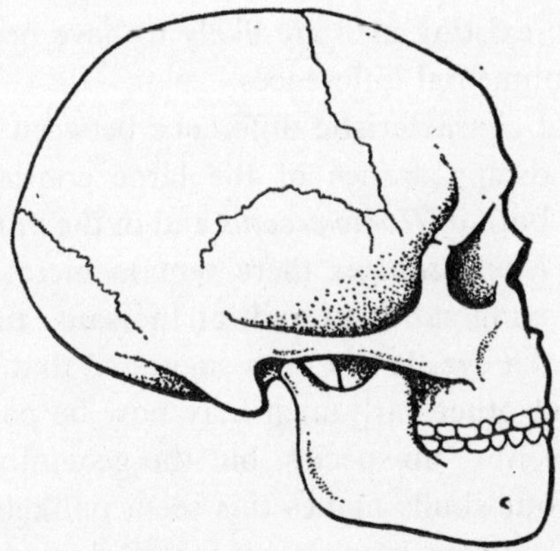

FIG. 85. Skull of Cro-Magnon type found at Combe-Capelle. (From an illustration in the British Museum (Natural History), by permission of the Trustees.)

The two types of people were separated from each other by a time interval of a few thousand years, the Combe-Capelle people arriving about 37,000 years ago, during the warmer interval of the first period of the Last Glaciation, whereas the Cro-Magnons appeared about 30,000 years ago, when the climate had turned cold again.

Until recently the distinction between Combe-Capelle and the Cro-Magnon has been specially emphasized. Now these differences are regarded as of less importance. Professor W. Howells has suggested that in structural features these two peoples might not have differed more than a modern Irishman and an Austrian.

Because the similarities shown by all European fossil men dating back to the same time are so pronounced, most anthropologists consider them all to have belonged to the Cro-Magnon type. The special importance of the Combe-Capelle fossils is that they mark the entrance into Europe of our own sub-species, now generally described as *Homo sapiens*. These early Europeans had only a few characteristics which would serve to distinguish them from some modern western and north Europeans, and there seems no reason to doubt that present-day European populations are in part descended from people similar to those whose skeletons have been discovered in the Upper Palaeolithic sites.

Most important, and relevant to a later discussion on race, is the realization that all living men belong to this one sub-species only, and none more so than others. Those obvious differences in stature and skin colour discernible in existing men are likely to have been an evolutionary adaptation to environmental differences.

Perhaps the most characteristic difference between *Homo sapiens* and earlier men is the disappearance of the large continuous supraorbital ridges so prominent both in *Homo erectus* and in the much later Solo Man. In our sub-species *Homo sapiens* there remain mere vestiges, the bony brows themselves having subsided and, at the same time, separated into two portions over each eye. It has been suggested that the beetling brows of *Homo erectus* and other early men may now be partly hidden by the vertical foreheads of our sub-species, but the generally lighter and more dainty structure of our skulls makes this seem unlikely. Apart from this principal difference, modern man's skull is well domed with vertical sides and a rounded brain case at the back.

These, then, are some of the features wherein our skulls differ from those of our ancestors. All very important no doubt from the anthropologist's point of view, but it still does not tell us where *Homo sapiens* actually came from. Some have attempted to answer the question in an easy way by allowing the Neanderthals to evolve straight into *Homo sapiens*. But this solution seems most unlikely, for the Upper Palaeolithic men were markedly different people from the Neanderthals of the Last Glaciation, and there do not appear to have been any intermediate types in existence in Europe. Others hold the view that Upper Palaeolithic men drifted into Europe, came into contact with the Neanderthals, and were actually able to breed with them. This solution is invalidated for the same reason as the 'gradual evolution theory'—intermediate types of hybrids of modern men and the Neanderthals having never been discovered in Europe.

There is, as mentioned earlier, remarkably little variation between the many skulls of the Upper Palaeolithic peoples of Europe (Cro-Magnon type) with only a few definitely distinguishable into the long-skulled Combe-Capelle people. Two exceptions of this trend have been discovered. One was found at Chancelade, in France, in late Upper Palaeolithic deposits. The skeleton of this man appears short in stature and the skull suggests certain Eskimo characteristics. The second unusual find appeared in the form of a pair of skeletons, a part of the now famous fossil finds from the Grimaldi caves carved out of a high cliff near Menton on the Riviera. There, from the lowest level of the Grotte des Enfants, a pair of skeletons, probably a mother and her child, were found buried together. Because of what he believed were certain negroid characteristics, R. Verneau named them the Grimaldi Negroids. In the absence of evidence that Negro-type humans lived in this part of Europe, these skeletons have posed an anthropological puzzle. It has been suggested that the skulls were badly reconstructed, but some believe that it would be unwise to dismiss these possible Negroes as an anthropological error because they appeared in an unlikely place. Better to leave the matter in abeyance until a simple and straightforward explanation for their presence can be found.

A Problem to be Solved

WE ARE STILL LEFT very much in the dark as to the origins of *Homo sapiens*. While most authorities favour the view that the last of the Neanderthals were gradually replaced by the Cro-Magnon people and their allied types, evidence as to our origin is still too thin to allow more than speculation on this point.

Now that we can state with some certainty that *Homo sapiens* did not evolve in Europe straight from Neanderthal Man, the puzzle to be solved relates to his and his predecessor's whereabouts during the Würm Glaciation, before his arrival in Europe. As far as the numerous types of *Homo erectus* are concerned, we have become acquainted with them in some detail but have so far encountered no evidence that during their time, half a million years ago, other types akin to the Neanderthals or Cro-Magnons existed. This leaves a distressing gap of nearly 400,000 years during which we know practically nothing about man's further progress towards the *Homo sapiens* state.

Judging from their tool industries the men of the Mindel and Riss Glaciations chipped away at their Abbevillian and Acheulian hand axes without making very much progress towards more sophisticated artifacts. It could be readily assumed that man's advances in tool-making also reflect the evolution of his brain. This may be a reliable guide but we cannot be certain. The man who left a footprint in the sands at Terra Amata on the French Riviera may have been a little over five feet tall if one uses the formula applied to Neanderthal footprints found in the Grotto of Toirano in Italy. In fact, judging by the height, he might have been a Neanderthal except that he lived 180,000 years before the first Neanderthals were known, and the tools he made appear too primitive for Neanderthal Man. Nevertheless he was not only a tool-maker but also a shelter-builder, and if he had only buried his dead where he had built his shelters and cooked his food, we might be the wiser now as to his appearance. This then does not help us very much in our search for our more direct ancestors. It does, however, indicate that men of considerable intelligence frequented Europe 300,000 years ago.

Let us now come nearer to the present and accompany a joint expedition of the American School of Prehistoric Research and the British School of Archaeology in Jerusalem in their excavations on the western slopes of Mount Carmel in Palestine. Here, as mentioned before, in 1931 and 1932 skeletal remains of at least a dozen people were discovered in the cave of Mugharet et Tabūn and in the rock shelter Mugharet es-Skhūl. The Tabūn site yielded one nearly complete skeleton of an adult woman who was very like the Neanderthal fossils of Europe, not an altogether surprising find, since the site contained a sequence of layers running from late Acheulian through Upper Mousterian stone tool industries.

A surprise awaited the archaeologists when the numerous skeletons in the Skhūl rock shelter were evaluated. The anatomical differences among the inhabitants of the two sites were immediately noticeable, for the Skhūl people were rather unlike the heavily built Neanderthals of the cold north-west, being tall, straight-limbed and approaching closely to *Homo sapiens* in appearance. Like ourselves in size and shape, their skulls were high and flat-sided, not projecting in the rear like the Neanderthal skulls. Yet the front of the skulls, particularly the brows, shows slight Neanderthal similarities. Thus the Mount Carmel fossils are one of the very few examples in human palaeontology in which some of the varying characteristics of a population can be observed.

Carbon-14 datings of the Tabūn deposit which contains the female skeleton indicate an age of 41,000 years. This makes it contemporary with the Neanderthals of Western Europe. The Skhūl population lived about 10,000 years later when the climate had changed and many animals, including the hippopotamus and rhinoceros, had disappeared from the area.

A number of theories have been advanced to explain the origins of the Mount Carmel fossils and their possible relationships with the Cro-Magnons—the modern men of Europe. Some favour the view that the Tabūn woman represents the Neanderthal stage of evolution in the Middle East, and the Skhūl people their descendants in the course of evolving toward the modern Cro-Magnons of Europe. As far as their evolution towards modern men is concerned this remains a distinct possibility, the only problem is that we do not know *which* modern men, for it is now almost certain that they could not have evolved into the Cro-Magnons of Europe, simply because there just was not enough time. If we consider that the Tabūn woman lived about 41,000 years ago and the Skhūl people followed her 10,000 years later, it takes them to the beginning of the Upper Palaeolithic, about 30,000 years ago, when our probable ancestors, the Cro-Magnons, had already arrived in Europe. While the possibility cannot be ruled out that the Skhūl people evolved into modern men in a place so far unknown, the Cro-Magnons do not appear to be their direct descendants.

This unfortunately leaves our origins as obscure as ever, so let us search for other fossil men that may throw some light on this subject. In our search for the early existence of *Homo* with a more modern form of brain we first visit one of the terraces of the lower Thames Valley at Swanscombe in Kent. There in 1935, A. T. Marston, a London dentist, discovered the back and bottom of the now famous Swanscombe skull twenty-six feet below the surface. The following year, at a point twenty-three feet away from the position of the first find, the enthusiastic dentist discovered the left side wall and top (parietal bone) of the same skull. Then, twenty years later, in July 1955, the same geological level yielded the right parietal bone of the same skull. Also found with the skull bones were flint implements which can with certainty be assigned to the Middle Acheulian hand-axe industry. These skull fragments have now been dated, suggesting an age of about 200,000 years. This antiquity is of particular interest because the woman of Swanscombe had a brain case of nearly modern size, with an estimated cranial capacity between 1,325 and 1,350 cubic centimetres. Unfortunately the front part of the skull was never found so we know nothing about the state of her brow ridges and the shape of her forehead.

In 1933 two years before the first Swanscombe discovery, a Dr Berckhemer found the skull of a woman twenty feet down in a gravel pit at Steinheim in Southern Germany. It lay in gravels of the Second Inter-Glacial Period together with the remains of extinct mammals. She was probably therefore a contemporary of the lady of Swanscombe. The skull has a rather low cranial capacity—about 1,100 cubic centimetres—and the brow ridges appear strongly developed, but the skull as a whole permits the inference that it could have belonged to a primitive *Homo sapiens* type.

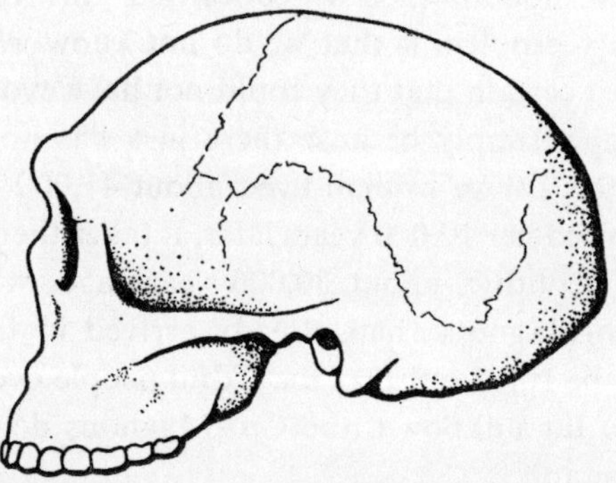

FIG. 86. The Steinheim skull seen from the side (partly restored). (From an illustration in the British Museum (Natural History), by permission of the Trustees.)

Thus the Steinheim and Swanscombe evidence suggests that by the Second Inter-Glacial natural selection had evolved brain cases similar to those of modern man at a time when the species *Homo erectus* must still have been much in evidence.

Were these perhaps the people who, 300,000 years ago, spent their summers at Terra Amata on the French Riviera, hunting and shelter-building?

More Ancient Skulls, with Modern Features

In 1947 the French archaeologist Mademoiselle Germaine Henri-Martin, while investigating the hard stalagmite base of the Fontechevade Cave in Western France, found among animal fossil deposits of the Third Inter-Glacial Period the greater part of the vault of one skull and a small piece from the brow region of another. Measurements show that this man had a brain case exceeding 1,450 cubic centimetres, and that it was in every way comparable to our own. More sensational was the evidence that the brows of these people could not be of the heavy beetling kind characteristic of all Neanderthal skulls, early and late. In fact both fragments indicate that the owners had a smooth forehead similar to our own.

These finds suggest that *Homo sapiens* had forerunners of modern skull form who were present in Europe at least as early as the first Neanderthals.

Similar evidence also comes from the African continent and was brought to light by Dr Leakey who, in 1932, found a number of fragments belonging to four skulls at Kanjera on the eastern shore of Lake Victoria. They show many similarities to *Homo sapiens* and none of the primitive traits of the Neanderthals. Dr Leakey found these fossils in association with the remains of a number of extinct animals, and after Dr K. Oakley had submitted these fossils to a uranium estimation test it could be confirmed that the animal and human remains belonged to the same period. Some doubt still lingers about the true age of the Kanjera fossils, but they appear to represent a modern type of man who lived in East Africa 60–70,000 years ago during a time when the Neanderthals were already prominent in Europe.

It seems, therefore, that some time during the later half of the Pleistocene, some half a million years ago, a number of forms of *Homo erectus* went their separate evolutionary ways leading, in one case, to the final enlargement of the brain in the Neanderthals, and in a second case, to an enlargement of the brain and a more general change of skull form in *Homo sapiens.*

When all is said and done, the problem of a cradle for *Homo sapiens* still remains and is further complicated by nearly as many theories about the subject as there are leading anthropologists. One of the most plausible views comes from Professor William Howells, who believes that *Homo sapiens* went on a separate evolutionary way some time during the later Pleistocene evolving from a primitive Neanderthal stock that also gave rise in another direction to the typical cave Neanderthals of Europe.

Rhodesian Man, a Probable Contemporary of the Cro-Magnons

DURING THE TIME when the Cro-Magnons had already arrived in Europe, 25–30,000 years ago, men of more primitive appearance inhabited parts of Africa. The most interesting and striking example is 'Rhodesian Man' who lived at the beginning of the Upper Palaeolithic in what is now Zambia.

There, in 1921, in the course of mining operations in the Broken Hill mine a short distance north of the Zambesi River, the Swiss miner Zwigelaar unearthed a skull right at the end of an ancient cave. Fortunately for posterity, this skull was well preserved and revealed valuable information about a new type of man who had lived in Africa. The skull is of unusually massive appearance with huge brow ridges, a strongly retreating forehead, a voluminous upper jaw and a wide palate. In some of these aspects it resembles Neanderthal Man but in others, such as the nose and ear parts of the skull, it appears different. Rather large in its overall measurements, it has a bigger and more primitive face than ours, though not a very projecting one. It is not as thick-boned as those of the ancient men of Java and Peking and its brain had attained the volume of about 1,300 cubic centimetres, nearly in line with modern man.

In contrast with the unusual appearance of the Rhodesian Man's skull the associated skeletal fragments are not distinguishable from those of *Homo sapiens* and show none of the features commonly regarded as characteristic of Neanderthal Man. The animal bones found in the cave and the relatively fresh appearance of the skull suggests that their age may not be very great, and artifacts found with these remains indicate a late Pleistocene date, estimated on indirect evidence to have been about 25,000 years.

For some time this skull remained somewhat of a mystery in archaeological circles, until in 1953 Dr Ronald Singer and Mr Keith Jolly of Cape Town University found further skull fragments at Saldanha in Cape Province, which when reconstructed appeared surprisingly similar to the Broken Hill brain case. The cranial capacity of this specimen has been estimated at about 1,250 cubic centimetres with similar huge brow ridges to his Broken Hill relative's. From the associated animal fauna and some late Acheulian tools it has been inferred that the Saldanha woman is about 40,000 years old, and thus of greater antiquity than her Rhodesian near-relative. Although scant in numbers these African fossils at least allow us a glimpse into a period of African prehistory when Europe was populated by the very last of the Neanderthals and the modern Cro-Magnons were beginning to arrive.

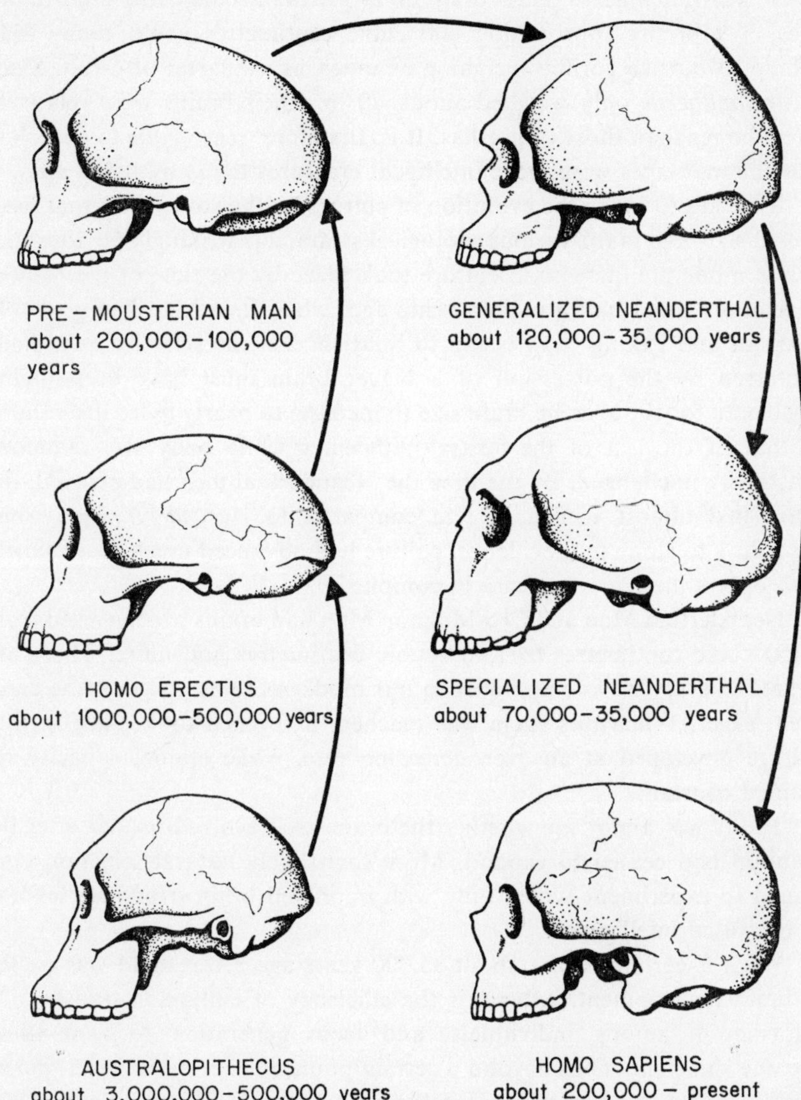

FIG. 87. Diagram illustrating the appearance of the skull in a series of fossil hominids arranged in approximate sequences. (From an illustration in the British Museum (Natural History), by permission of the Trustees.)

An Important Threshold

LET US NOW take a final look at the evolutionary progress of the hominid brain. *Australopithecus* had a brain no larger than some living anthropoid apes. Yet brains approaching 600 cubic centimetres could today only belong to outsize gorillas weighing as much as a quarter of a ton. Since *Australopithecus* only weighed about 80 lb., their brains were relatively much larger than those of gorillas. It is, therefore, reasonable to speculate that the man-apes were more intelligent creatures than any living apes.

If we are to judge the evolution of culture by the stone tools that have been discovered in the various geological strata, it is strikingly obvious that only a moderate increase in culture took place by the time of the Mindel Glaciation about half a million years ago, when Java Man had probably died out and Peking Man began to flourish. Yet the selective advantage conferred by the possession of a bigger brain must have been highly significant for the average brain size to increase to nearly twice the volume of that of the last of the australopithecines while body size remained practically unchanged. By the time the Neanderthal men had evolved, the brain had almost tripled in size compared to *Australopithecus*, while according to stone tool evidence, culture had advanced much more slowly with only a moderate increase in complexity.

Neanderthal Man and Cro-Magnon Man had brains which varied from 1,450 cubic centimetres to 1,660 cubic centimetres and more. These are figures in every way comparable to our modern brains and in some cases even larger. When this stage was reached, a reversal took place in that culture developed at an ever-increasing rate, while cranial capacity remained constant.

Today we cannot say whether the brain itself evolved further after the cranium had ceased to expand. More than likely natural selection continued to experiment successfully with improved brain structures leading to increased intelligence.

What is certain is that about 35,000 years ago a threshold was passed in brain development. After this the efficiency of culture in transmitting information among individuals, and from generation to generation, became sufficient. Thus beyond a certain point, further brain development leading to increased individual intelligence conferred no particular survival advantage upon the possessors. This does not mean, however, that the biological evolution of man stopped when culture appeared. Although modified, it is still continuing today.

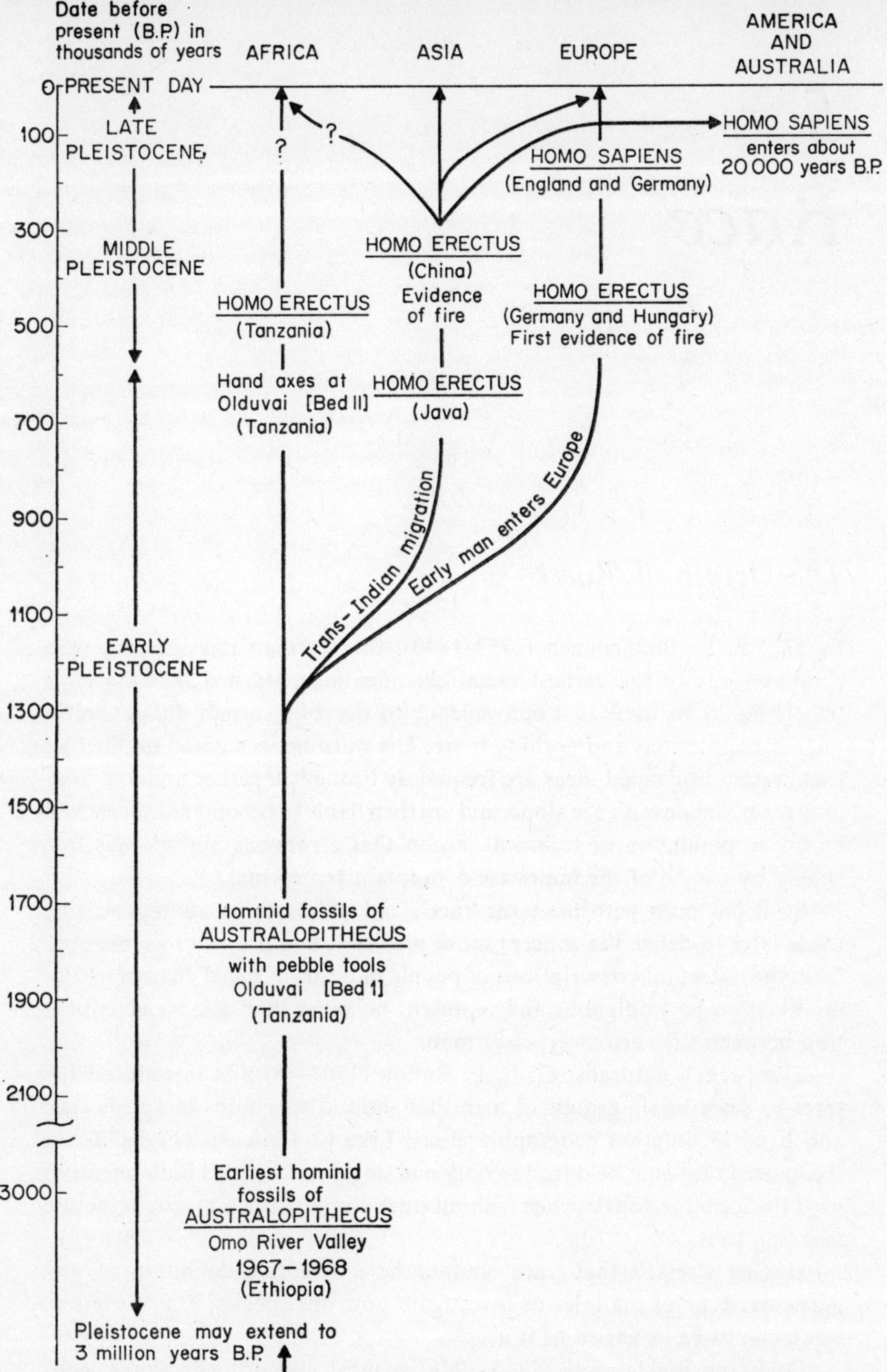

FIG. 88. This chart indicates approximately some of the dated archaeological discoveries from the Pleistocene and their possible relationships in terms of the migrations of human ancestral groups. From at least one million years ago migrations between Africa, Europe and Asia were probably continuous, and no attempt has been made to show the complexity of these movements. (Adapted from a drawing by Bernard Campbell.)

5
Race

The Origin of Races

In 1775 J. F. Blumenbach (1752–1840), the German physiologist, who proposed one of the earliest racial classifications, warned that the term 'race' should be used as a convenience to describe certain differences in human populations and nothing more. His warning was based on the fact that certain ill-defined ideas are frequently brought together under a common term for convenience alone, and are then liable to become so entrenched by an accumulation of technical jargon that erroneous authority is lent simply by reason of the imprecise concepts it represents.

So it has been with the term 'race', and although an attempt will be made later to define the concept more precisely, it is prudent to remember from the outset that descriptions of people in terms of racial characteristics are likely to be ambiguous and represent no more than a loose classification between the various types of man.

The French naturalist G. L. L. Buffon (1707–88) first introduced the term to describe six groups of men that showed variations in appearance and lived in different geographic areas. Like C. Linnaeus (1707–78), he recognized that man belonged to only one single species, and both scientists used the term for convenience without intending to attach precise scientific meaning to it.

Having stressed that 'race' cannot be a scientific definition of any particular type of man, let us investigate how our species, *Homo sapiens*, has come to be as varied as it is.

Dr Franz Weidenreich (1873–1948), a most able anthropologist, conceived a theory suggesting that the modern races of man descended from four principal ancestral lines, viz. the Australian Aborigines from Solo Man and Java Man; the Mongoloid strains from Peking Man; the African Negroes from Rhodesian Man, and the Eurasians from the Tabūn people, and later, from Cro-Magnon Man. This theory, however, although supported by a number of other distinguished workers, suffers from two distinct weaknesses. First, it assumes that none of the lines of ancient men enumerated above were caught in an evolutionary blind alley and died out, a possible but unlikely proposition. Second, many more marked distinctions exist between these fossil men than between the living races.

An increasingly plausible suggestion is that all human varieties living at present derived from a relatively small group of early men of the *Homo sapiens* type, originating perhaps half a million years ago. As this small but variable group increased numerically, and segments moved into different climatic and geographical regions, genetic isolation and differential selection probably led to the formation of numerous population strains. These, in turn, gave rise to further races down to the present time. With continuous changes taking place, clearly no contemporary population exactly resembles the original *Homo sapiens* men, nor can it be claimed that any particular races are purer or less changed than others. Nor can an examination of existing races determine what the skin colour, hair form, or blood groups of the earliest races of man might have been.

The causes of racial variations can be accounted for by a number of important factors, the most important being natural selection, genetic drift, interbreeding and mutations.

Natural selection whittles away less adaptive traits and favours more adaptive ones under the particular circumstances, working towards a genetic constitution best fitted to prevailing conditions.

In any population the gene frequencies change in a random manner from generation to generation even though the environmental conditions may remain the same. This random fluctuation of gene frequencies, known as genetic drift, was first studied by Sewall Wright. It arises from the fact that each generation is a result of the union of a random sample of gametes produced by the parent generation. The smaller the sample of gametes, the larger is the sampling fluctuation in the gene frequencies. Or a small migrant group may branch off a larger group and, by chance, have no individuals with a particular gene. Again, such a small isolated group might have only a few carriers of this particular gene, and none of these genes are passed on to succeeding generations. In this way it is likely that blood group B was lost by the small isolated group of Arctic Eskimoes of North Greenland.

Interbreeding between races results in new gene combinations, and is one of the most common ways in which gene frequencies are altered.

Genetic drift and natural selection were probably of great importance as race-forming mechanisms in past ages when man still lived in small isolated groups. Indeed, natural selection is likely to have been among the most important forces which brought about the differences which we discern among racial groups. Although a racial feature like skin colour appears to be a direct adaptation to climate, closer examination does not in all cases lead to straightforward explanations as to how such features might have been acquired. On the other hand, it has been known for a long time that the body build of man, and that of warm-blooded animals in general, is closely related to the climates of their habitats. For example, an animal species that occurs in widely separated areas with different climates is normally found to be somewhat larger in the colder parts of its range. This general principle also appears to apply to man. The reason for this variation in stature is closely connected with body heat preservation and body heat loss. In a cold climate an animal will require an efficient mechanism to retain the heat generated by its metabolism. Similarly in very hot climates the reverse applies, making for the kind of body form best suited for dissipating excess heat. Thus, the ratio of heat produced against that dissipated is greater in more bulky individuals. Biologists noted also that those parts of the body that present an extensive surface to the air, noses, ears and limbs, tend to be smaller in colder parts of an animal's range than in warmer parts.

So far as our species is concerned, we also seem to have evolved along similar patterns. Thus, Indians inhabiting Canada, North America and the Southern parts of the South American Continent tend in general to be larger than their cousins living in warmer climates. It is also clear that the peoples of hot desert climates are generally long-limbed and lanky. Short-limbed, stout people occupy the Arctic regions.

This seemingly universal trend among warm-blooded mammals, however, does not help very much in our search for the origin of racial characteristics. As mentioned previously, it is a trend found among most races, and dependent upon the climate of their home areas.

Of all outward human racial characteristics, skin colour is the most marked. The pigmentation in dark-skinned people is due to an organic molecule called melanin. Its exact function is still being investigated, but there is evidence that melanin in the outer layer of skin can absorb the harmful ultra-violet radiation from the sun, thus preventing it from penetrating and damaging the living cells of the skin. Present in all humans except albinos, our bodies seem to be capable of manufacturing it in varying amounts, depending on our genetic constitution and the measure of sunlight to which we are exposed. Thus races differ in the basic amount of melanin present in their skins and this is determined by their genes.

References to a 'white' and 'black' race are, of course, inaccurate and misleading, for human strains vary in skin colour from very pale, as in northern Europeans, to nearly black-skinned peoples inhabiting parts of Africa. In between the two extremes are many brown-skinned tints, but it would be wrong to jump to the conclusion that these have resulted as an adaptation to the varying strengths of sunlight found in different parts of the world. It is reasonable to argue that interbreeding between racial groups of differing skin colours must, in part, be responsible for the many basic variations in the melanin content of human skin.

It requires to be noted, however, that dark-skinned human beings are most often to be found living in regions of intense sunshine of the Old World, while light-skinned Europeans originally lived in regions of the Northern Hemisphere where the ultra-violet component of solar radiation is less strong.

Limited amounts of ultra-violet light are needed in the formation of vitamin D and it can be argued that in areas of weaker solar radiation, such as Northern Europe, natural selection has favoured individuals with less melanin in their skins simply because it enabled their bodies to manufacture more adequate quantities of vitamin D, probably making them fitter.

All this may sound reasonable, but the racial jig-saw puzzle based on colour does not quite fit in every aspect.

While it is probably true that natural selection favoured people with dark skins living in areas of strong solar radiation, such as the equatorial regions in Africa and the Western Pacific, it is also a fact that large parts of Negro homeland consist of forests where bright sunshine can be completely absent.

For example, the Pygmies mostly inhabit forest areas in Africa and South-Eastern Asia. The desert habitats of Northern Africa are inhabited by peoples of slim and tall body form, as might be expected in a hot climate, yet their skin colours vary widely from a light tan to almost black.

Skin colour, obviously, cannot satisfactorily be explained by tying it directly to the amount of sunlight available in the various regions of the world. It is more likely to be the result of a complex balance of influences which includes other factors. There are indications that densely pigmented skins have an increased resistance to infection. It has been suggested that a dark human skin served as a protective colouration in the case of forest hunters. This could have worked in two ways, by camouflaging the hunter from his prey, and also protecting him from predators.

Another possibility is that the dark-skinned peoples of Africa are newcomers to the forest areas, having evolved their skin colour in a sunny, more or less treeless habitat like the Southern Sahara. The Pygmies who now live in forest areas may be representatives of such people who took to a forest existence after natural selection had endowed them with their dark skins. One might also speculate that their small stature is a useful adaptation in densely forested regions, for it clearly makes for greater mobility and speed in the dense undergrowth of the African jungle.

The fact that different races inhabit widely varying climatic zones, ranging from Arctic to tropical, suggests some degree of genetic adaptation. There is, indeed, indirect evidence of this in the fact that, for example, the people of Tierra del Fuego, the archipelago at the southern extremity of South America, as well as some Australian Aborigines, are able to sleep unclothed at near freezing temperatures because their metabolic rates and skin temperatures are lowered. Furthermore, evidence showed that American Negro troops experienced greater discomfort from cold during the Korean war than did white soldiers, who also withstood humid heat better than the black-skinned troops.

Experimental studies show that Eskimo peripheral skin temperatures remain higher than normal in the cold, enabling them to retain the finer motor skills in their fingers when their hands are exposed to very low temperatures. While some of these adaptations appear to be tied to specific racial groups, it is as well to remember that these differences point more to the importance of local population groups and local environmental factors. Thus, for example, Australian Aborigines of the tropical zones appear to lack the cold adaptation of their southern brothers.

As is well known, the hair and eye colour of man varies almost as much as that of his skin. But the melanin granules within the hair are arranged in such a manner that only a moderate amount of the pigment will make it appear black. This is the reason why a good many light-skinned people possess dark hair. Only where de-pigmentation has become more complete do we normally get individuals with light brown or blonde hair. Both hair and eye colour therefore appear to have little selective advantage, and are probably simply a side effect reflecting the melanin contents of the skin.

It is immediately obvious that hair varies not only in colour but also in shape. Generally we associate yellow-skinned peoples with perfectly straight, long and rather coarse hair, dark brown or black in colour with a nearly round cross-section, viz. the Chinese and the American Indians. Negroes, on the other hand, have frizzy, woolly and peppercorn hair nearly always dark brown or black, rather short, elliptical or kidney shape in section. It grows with a continuous curl in all the hairs, which do not lie parallel, thus forming a vigorously intertwining surface. Wavy and curly hair, rather smooth and silky, oval in section, is typical of Europeans; sometimes the hairs twist in unison at intervals along their length.

Unlike hair colour, hair *form* does appear to have definite adaptive significance. The head, the container of the vital organ, the brain, is vulnerable to blows and to overheating from direct solar radiation.The insulation provided by densely intertwining Negro hairs and that of allied human types must therefore be considerable. Thus the shape of hair, like skin colour, probably evolved in response to selective pressures.

Other racial features include characteristics such as the thick lips of the Negroes, the full beards of the whites and Australian Aborigines, who have also retained more body hair than other varieties of man. Skull shapes also vary considerably among different human strains. None of these racial features can be readily explained, and they are probably the result of causes no longer operative.

Many varieties of the human species are identified by rather loose groupings, such as the Pygmies and the tall Watutsi of Africa, the Polynesians of the Pacific areas, the dark-eyed Mediterranean people, the light-skinned Europeans, the Bushmen of Africa, the Indians of the Americas, the Mongoloid peoples of China and Japan, and others. These names, however, are only useful as a general kind of reference and have little biological significance.

Anthropologists have tried to group a number of similar populations into a 'race'. These races have been described in terms of such external features as skin colour, width of nose and hair shape. Using these characters, the Caucasoid race was described as wavy-haired, narrow-nosed and white-skinned; the Mongoloids as straight-haired, moderately broad-nosed and yellow-brown-skinned; the Australoids as curly-haired, moderately broad-nosed and brown-skinned; the Negroids as woolly-haired, broad-nosed and black-skinned. These have become popular and generally accepted descriptions of the chief races of the world.

Closer examination, however, shows a number of serious inadequacies springing from this superficial classification. If the noses, skin and hair of men were to be ignored, it would be extremely difficult to tell one racial type from another. Even in these features there is much general overlapping which makes the lines of distinction very vague. Many anthropologists have attempted to classify man by external physical characteristics, and some have even claimed they can classify skull forms, but the difficulty remains that in all groups there are always many exceptions.

It was not until early this century that the A-B-O blood groups, the Rh blood types and others were discovered and began to attract the attention of anthropologists. Genetically determined, these blood groups are universally distributed among all mankind. Later it was realized that the frequency of types A, B, AB and O blood differed in different populations. Further studies of blood type distribution patterns, now based on sound biological evidence, have been invaluable in determining some of the relationships of human populations.

Biologists soon found that different populations varied widely with respect to these traits. For example, it was discovered that type B blood was absent among American Indians and the Australian Aborigines, and was very rare among the Basques of Northern Spain. When their blood was typed for the Rh factor, it became evident that the Basques possessed the highest proportion of Rh negative persons in the world, whereas the Australian Aborigines, American Indians, Indonesians and Thais included hardly any Rh negative individuals. Furthermore, while blood type A is found in all geographical races, the A2 sub-type is uniquely European.

This and other data suggested that populations in different parts of the world differ widely in the proportions of the blood group alleles they possess. By preparing special blood gene frequency maps of the peoples of the world, much useful information on human populations has been gained. For example, such a study revealed that the frequency of the allele for type B blood is very high, exceeding 25 per cent, in extreme Eastern Europe, and higher still in Tibet and parts of India. Moving westward it gradually decreases in all directions until it falls below 5 per cent in parts of Scandinavia, France, Spain and Portugal.

One explanation for this gene gradient found in Europe has been that type B alleles were carried by Asiatic peoples migrating westward. They brought their type B alleles to the gene pools of the peoples already inhabiting Europe.

Why the type B blood group should be so common in Central Asia is not known, but it is possible that this blood group may be linked to other physical attributes of man which had a distinct survival advantage in a particular environment but now are no longer apparent. In those areas where type B alleles are still rare, it would be reasonable to assume that very little gene mixing occurred.

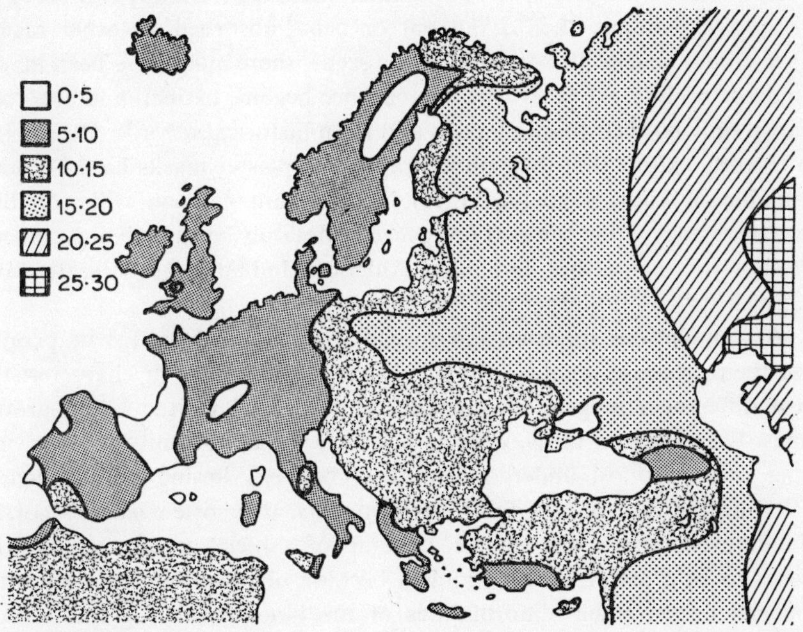

FIG. 89. The frequencies of blood group gene B in different populations of Europe. (Adapted from Biological Sciences Curriculum Study, *Biological Science: Molecules to Man*, Houghton Mifflin Co.)

These and similar studies revealed factors of great biological and anthropological interest, but apart from the scientific side of the story, these studies also indicate that the racist doctrine of a master race, whichever colour it might be, is completely unsupported by emerging facts relating to the genetic make-up of man.

Any description of human populations in terms of gene frequencies will diminish in value as time progresses. Races are no longer isolated from each other by geographical barriers. Modern systems of communication enable many more people than formerly to travel around the world and make contact with each other, thereby increasing social contact between all types of people and resulting in further mixing of the human gene pool.

South American countries, the United States, and more recently Europe, are becoming multi-racial societies, and it is to be expected that more gene exchanges between the peoples of the world will cause differences among populations to lessen.

In the history of any species of animal, races are normally transitory—forming, expanding, then dying out or being absorbed by other races. Since *Homo sapiens* first arrived on the scene, there must have been many hundreds of different races which have since become extinct, and are now known only by their fossil remains and from historical records. Others are so recently extinct that their photographs and plaster masks have become available for study. Some local races, like the Ainu of Japan, will probably become extinct within the next few decades, not only because their members are slowly dying out but, like many American Indian local races, they are being absorbed into larger communities.

As populations increase and spread, smaller local groups of people lose their separate identity, though leaving their traces in characteristic local differences in gene frequencies, as evidenced by the blood group genes. Even isolated races, who do not have the opportunity to mix with other racial groups, undergo continual change. Altering environments, including the disease and food environments, may often lead to corresponding changes in the genetic make-up of a racial group. As malaria ceases to be a major health problem because of insect control and anti-malarial drugs, some abnormalities of the blood rendering individuals resistant to the disease will probably decrease in frequency until, ultimately, they may be as rare in the tropics as they are now in Northern Europe.

Neither will the races of today continue for ever, since they are contributing to new ones in process of formation. It is likely, however, that in a few thousand years from now *Homo sapiens* will have become a much more homogeneous species than it is today.

Taking a broad view of the racial differences to be found among the peoples of the world, it becomes increasingly clear that they resulted from a number of factors. Some are probably left-overs from certain evolutionary paths which are no longer obvious. Examples of these may be the reduction of brow ridges in *Homo sapiens* and the partial disappearance of body hair. Others, like skin colour and hair form, appear to be adaptive.

It is also apparent that there are many population groups where individuals share similar sets of genes, which in part manifest themselves in external features such as skin colour and hair forms. Such sets of genes, however, are only similar and not identical, and no single individual can be typed as an ideal member of a race.

The modern concept of race based on genetic studies of human populations claims that race differences are compounded of the same genetic variations in which individuals within a race also differ. The distinguished geneticist Professor T. Dobzhansky defines races as populations that differ in the frequency and prevalence of certain genes. These differences are also relative, for an individual may belong to more than one race or to no race at all. Thus a person may possess a combination of genes which are characteristic of different race populations. *Race* is, therefore, a transitional concept which describes rather vaguely populations with similar gene frequencies. On the other hand, *species* is a much better defined concept—at least in the sense that an individual belongs to one and only one species.

The Nature-Nurture Controversy

THE MOST SENSITIVE ASPECTS of the race controversy produce questions relating to alleged intelligence differences between the various races of man. At the centre of the argument stands the question asking which human traits are due to heredity (nature) and which to environment (nurture).

The idea that environment moulds us into what we are is a very old one, and was already stated by the English philosopher John Locke (1632–1704) in 1690. It found many followers during the eighteenth century, and was further popularized by Thomas Jefferson (1743–1826), third president of the United States, together with some of his contemporaries. Locke postulated that human beings are born with neither a good nor a bad character. The attributes and qualities which will appear during their upbringing are dictated entirely by the environment in which they find themselves. Thus according to Locke, a good education and a favourable home environment will introduce good potentialities into a person's character. This idea forms part of the basis of democracy, but if carried too far ends in an attempt to make people as much alike and uniform as possible by prescribing the same type of education for everyone irrespective of the individual capabilities of each person.

On the other hand, the basis for the hereditarian view was already laid by Aristotle when he expounded his dictum: 'Those who are sprung from better ancestors are likely to be better men, for nobility is excellence of family.' Followers of this view believe that a person's ability and character are essentially settled and predetermined by his heredity. Certain religious teaching gives support to this view in the dogma of Adam's original sin—inherited by all mankind to give human nature a basically evil slant. Calvinist tradition dictates that some humans are elect and others are damned by God's unalterable decision.

These and similar beliefs eventually gave rise to certain doctrines of convenience. For example, a person's wealth and position in society may also be predetermined by his hereditary make-up. It is a short step from there to the Nazis' racist policies with their evil results, and to many of the discriminatory racial attitudes so widespread all over the world today. From a political point of view one of the greatest dangers in the hereditarian theory lies in the assumption that man can do little to alter the lot of the under-developed peoples of the world because their heredity has dictated their present condition. An immediate reaction to this line of thought will be made by some people who feel that today's 'enlightened society' would hardly follow such naïve and uninformed doctrines. Yet it is a fact that it is this doctrine that underlies many of the racial prejudices found in Great Britain and other countries of the world.

What then is the answer to the nature and nurture problem? The truth seems to lie somewhere between the two, but it would be unwise to try to divide human traits into precise hereditary and environmental pigeonholes. Most human characteristics are influenced and modified both by a person's heredity and the environment in which he lives. In any given characteristic, however, either heredity or environment may predominate. A simple example can illustrate this point. An active sunbather will, after a while, develop a deep skin tan—the environment has exerted its influence—yet this same person's skin pigmentation is influenced also by heredity.

No conclusive answers to the nature-nurture problem are as yet available, although hereditarians and environmentalists have both put forward theories. What conclusions, therefore, can be drawn from them?

The hereditarians attempt to show that races cannot possess equal intelligence because they appear to differ in average intelligence as measured by I.Q. test performances. They further support their argument by the fact that different racial groups vary in brain size.

Taking first the question of brain size, there seems no reason whatsoever to believe that intelligence or mental capacity is simply proportional to the size of the brain.

We have seen that the Neanderthals had brains, on average, larger than human populations today, yet it seems unlikely that they were more intelligent than our sub-species, for they became extinct through possible competition with other men who had smaller brains. Men also have on average larger brains than women and, while some men might like to use this as an argument for the superior intelligence of their sex, there is no scientific evidence to support this idea. As to the races, they do differ in the average cranial capacity, but there are individuals in every race with much larger and much smaller brains than the average for any other race. Brain size is also correlated with average body size. Whale brains are much bigger than human brains, yet this does not mean that whales are far more intelligent than men. Also smaller peoples living in warmer areas of Europe, Asia, America and Africa have cranial capacities which average between 1,300 and 1,400 cubic centimetres, whilst some populations of Siberia, North America and Polynesia have the largest cranial capacity of any living peoples—1,500 cubic centimetres or more for adult males—yet these populations are not more intelligent than others with smaller brains. Anatole France, the famous French writer, had an estimated brain size of about 1,000 cubic centimetres and Ivan Turgenev, the Russian writer, had one of the largest brains, exceeding 2,000 cubic centimetres. Both were highly intelligent people, and the dimensions of their crania were obviously related to relative body size.

Intelligence tests do measure variable traits of some significance for academic potential, and these traits are in part controlled by genetic differences. Unfortunately, nobody has as yet succeeded in devising a culture-free intelligence test which would be equally suitable not only for people brought up in countries with different cultures, for example, Europeans, Japanese and Australian Aborigines, but also for those groups reared in the same country but having different social or linguistic backgrounds.

Clearly tests—quite reasonable within a certain culture and social class—may be unanswerable for people with different cultures. A Bushman may be highly skilled in interpreting the tracks of wild animals from insignificant signs on the ground, yet he may fail an aesthetic judgment test where it is required to choose between two similar drawings of the same subject for their artistic merit. The reverse would be true for a European.

Army intelligence tests carried out on Negro and white recruits during World War I showed that the mean scores for Negro draftees were lower than that for white ones.

These results were quickly interpreted by some as proof of Negro racial inferiority, but after the data had been properly evaluated, it became apparent that the mean score of the Negro draftees from Northern States was higher, not lower, than the mean score of the white draftees of the Southern States. Later studies, chiefly of schoolchildren, also showed that among the Negroes the mean I.Q. rose hand in hand with the duration of residence in Northern States, demonstrating that social and cultural environment was closely related to intellectual performance.

To date there is no evidence of the existence of genetically conditioned differences in mean intelligence between Negroes and whites, or for that matter between any races anywhere in the world. Of course this does not mean that such differences cannot exist. It merely suggests that at present no satisfactory tests are available to allow one to reach a scientifically valid conclusion.

One important aspect often overlooked, particularly by the culturally advanced peoples of the world, is that intelligence variations within any race, be it white or coloured, are much larger than variations of intelligence between races. Putting this another way, it means that the high I.Q.s of persons in every racial group are much higher than the averages for their own or any other group. And conversely the low variants in every race are much below the average for any race. In other words, men must be judged on their own merits as individuals. The disastrous prejudices which lead men of one colour to point at men of another, proclaiming them to be inferior because of the melanin content of their skin, is not only morally wrong but clearly without scientific foundation.

That environmental factors can strongly influence the development of intelligence has been shown by Dr H. H. Newman and his colleagues. Their studies revealed that human identical twins reared apart showed differences in I.Q. almost twice those obtained by identical twins reared together. In the 1950s a growing accumulation of research evidence began to point to the importance of early experience both before and after birth in the development of intelligence. The nature of the intervening mechanisms is not yet understood but climatic and nutritional factors appear to be of importance in the pre-natal period, whereas influences associated with premature birth and stimulus deprivation seem most powerful at birth and immediately after. At later age levels great attention is being paid to the role of social and motivational factors. For example, it is becoming increasingly clear that mental development as measured by intelligence tests varies markedly as a function of the family's cultural background and social and economic position. Nor does it appear likely that it will be possible to design culture-free intelligence tests. The evidence further suggests that along with inherited ability and the effects of training intelligence tests also reflect the value which the person places on intellectual activity and his general need to achieve or excel.

Despite the complexity of the subject and the clear limitations of current intelligence tests, the standardized techniques which they employ continue to be useful in such practical problems as the diagnosis of mental deficiency, the prediction of academic potential and success, and the selection of personnel in industry and in other occupations.

Race and Culture

THE QUESTION whether racial attributes have any concern with cultural achievement is an obvious one, for in the world today different races have achieved very different kinds of culture.

The exact causes for such differences are difficult to ascertain, but clearly cultures differ from one another to the extent to which the history and the experience of each of the interacting groups has differed. The distinguished anthropologist and sociologist Dr Ashley Montagu describes history in this sense as the experience that a person or group of persons has undergone or lived, perceived or sensed. No matter with what racial group we may be concerned, their kind of culture, whether it be that of a highly developed nation with a sophisticated technology, or a still simple Stone Age culture of a group of Australian Aborigines, is in its broadest and fundamental sense not merely an aspect, but a *function* of experience.

The reason why cultures of other racial groups often differ so much from our own is clearly explained by the fact that they have been exposed to an environment which differs as markedly from our own as do the cultures in question. If an Englishman, an Austrian or an Italian were to be brought up among a group of isolated Australian Aborigines, they would culturally become Australian Aborigines and behave like them, though physically they would remain members of their own racial group. The experience of a person, or groups of persons, is determined by their place of domicile and their cultural surroundings, and for this reason persons of similar intelligence belonging to different cultures may appear to differ mentally from one another. Thus Europeans brought up among Australian Aborigines would physically remain Europeans because their physical features are mostly genetically determined. But their physical expressions are likely to have been modified, and certainly they would have adopted the culture of Australian Aborigines. The reasons for this are fairly obvious because by culture we understand all those learned activities which we acquire from the society we live in, and this includes language, religious beliefs, institutions, customs and ideals. Unlike one's physical appearance, culture is acquired by learning and experience. Clearly our genes enable us to acquire culture, but they do not dictate to us which culture we can acquire. Therefore, the cultures of different societies and racial groups will differ according to the kinds of experience they have been subjected to.

Professor Erwin H. Ackerknecht has assembled data concerning white children who had been abducted from their parents by North American Indians during the eighteenth and nineteenth centuries.

Eight substantially reliable life histories of such abducted children have become available. The children were all between the ages of four and nine years, with the exception of a girl who was taken in adolescence. All of them appear to have adapted quickly to their new surroundings, forgetting their native culture to become completely Indianized. Later these individuals never attempted to return to their white relatives and to the culture of their birth. In the words of Ackerknecht: 'These "White Indians" seemed to have found a kind of unity of thought and action and a social cohesion which deeply appealed to them, and which they did not find with the whites, especially not with the pioneers.' There is no doubt that this fact largely contributed to their staying with the Indians. Dr Ashley Montagu stresses the fact that these 'White Indians' not only became completely Indianized culturally in the sense of manifesting purely Indian forms of social behaviour, but they also developed all the physical powers of resistance said to be peculiar to Indians. They also lived to an extremely old age, and the reports suggest that they had acquired the facial expressions characteristic of the Indian.

Reliable evidence of this sort is unfortunately not readily available, but those facts that could be established are sound evidence against the fallacy that culture is a genetically determined attribute which will express itself in an individual no matter what the environmental influences to which he is exposed.

This does not mean that individuals do not differ in those genes which influence temperament and intellectual capacity. Just as no two people have ever been found to have the same fingerprints, there are probably no two individuals who have identical intellectual ability. Comparisons between different societies and races as to their cultural abilities can only fairly be made if they have undergone very similar experiences throughout their histories. It also follows that cultural achievement is an exceedingly poor measure of the cultural potentialities of a person or a group.

The French statesman and economist Robert Jacques Turgot (1727–81) said very much the same thing when he wrote:

> The human mind contains everywhere the germs of the same progress, but nature partial in her gifts, has endowed certain minds with an abundance of talents which it has refused to others; circumstances develop these talents or relegate them to obscurity, and to the infinite variety of these circumstances is due the inequality in the progress of nations.

There are certain needs which are common to all people, but how these needs are satisfied depends very much on the traditions which have been adopted by the various human racial groups and their societies. For example, all of us have to eat, but what we eat will very much depend, in the first place, on the foods most readily available and on the traditions concerning its preparation, which are again largely culturally determined.

Our species possesses a genetic attribute which enables it to speak, but the language and mannerisms we adopt in our speech are determined by what we hear and see in the culture in which we have been brought up.

Tired people anywhere in the world experience a strong desire to sit down, to lie down, or to sleep, but the manner in which they take up their resting positions is culturally determined by the customs of the society in which they live. These examples are only a few of many culturally determined forms of behaviour. Their importance stems from the fact that they show that many of our basic biological needs are fulfilled through more or less culturally controlled processes, and the way we satisfy them is normally dictated by the traditions and social laws of the particular society to which we happen to belong.

To make judgments on the intelligence of racial or social groups without taking into consideration the cultural influences which have played a part in forming their behaviour patterns is both unscientific and nonsensical. Not only must we take into consideration a people's environment before judging their behaviour, but we must also give due consideration to their sets of values which are not only likely to be different from our own, but may in some aspects be superior.

John Stuart Mill (1806–73), the British philosopher and economist, said in his book *Principles of Political Economy*: 'Of all the vulgar modes of escaping from the consideration of the effect of social and moral influence on the human mind, the most vulgar is that of attributing the diversities of conduct in character to inherent natural differences.' Walter Bagehot (1826–77), the English economist and sociologist, wrote in his *Physics and Politics*: 'When a philosopher cannot account for anything in any other manner he boldly ascribes it to an occult quality in some race.'

Fortunately, an ever-increasing number of people all over the world are now dismissing such absurd ideas that cultures are irrevocably tied to the genetic constitution of races. Unfortunately, there are still a good many people in all walks of life who are convinced of the contrary; and even if not convinced, they would like to use these destructive notions for political and economic ends. Such men can be found among diverse political and religious persuasions, and while their actual numbers may not be large, they are often highly talented and vociferous and are able to hold sway over large sections of public opinion.

Man must begin to understand that all cultures, whether they are the product of black, white, or yellow peoples of the world, should be evaluated in relation to their own history and not by the chosen standard of any single culture, as, for example, that of Europe or the United States.

Cultures are the result of the responses to the conditions which a people have encountered throughout their history. If these conditions were limited in nature, the culture which has sprung from these will reflect this. If the conditions have been many and varied in character, then the culture will have assimilated much of that variety. Anthropologists and biologists generally agree that *Homo sapiens* as a species is adaptable and sensitive wherever it is to be found. Thus the most satisfactory explanation for the mental and cultural differences between the varieties of mankind is to be found in their different experiences.

Five thousand years ago, in what is now Great Britain, our ancestors had not advanced far beyond an Upper Palaeolithic Stone Age culture. Later the Roman statesman, Marcus Tullius Cicero (106–43 B.C.) had this to say about Britain: 'There is not a scrap of silver in the Island, nor any hope of booty except from slaves; but I don't fancy you will find any who has literary or musical talent among them.' Cicero must have been mistaken in his assumption, for the talents were there but it took some hundreds of years of development before they could find expression.

Many important discoveries on which our civilization rests, writing, the alphabet, the wheel, the astronomical calendar, and others, originated from the Middle East at a time when much of Europe was inhabited by tribes of so-called barbarians. It was, however, the descendants of those Europeans who, five thousand years ago, had a primitive culture, when the peoples of Egypt of the first and second dynasties (3200–2780 B.C.) already enjoyed a high degree of civilization, eventually caught up and in some aspects—such as the development of the sciences—surpassed the Middle Eastern civilizations.

Dr Ashley Montagu believes that the rate of cultural change is dependent on many different factors, but the indispensable condition for the production of cultural change is what he calls 'the irritability produced by the stimulus of new experience'.

Time, he says, is only a convenient framework from which to observe the development of culture or cultural events, and clearly many cultural changes, which among some societies have taken hundreds of years to come about, are among other peoples effected within decades.

Time seems not to be the most important factor in the development of cultures, but experience which springs from the variety of cultural contacts a people happen to possess.

The genes of racial groups do not, therefore, determine the types of culture they can acquire. The basic biological constitution for the acquisition of culture exists in all types of men, whether they be the Australian Aborigines who, until quite recently, had not advanced much further than a Stone Age culture, or the ancient Britons who, from Neolithic times onwards, had increasingly frequent contacts with the peoples of the continent of Europe, allowing the former to acquire a more sophisticated culture by the time the Romans landed on their shores more than 2,000 years ago. The Australian Aborigines, on the other hand, have until recently led a completely isolated existence on their continent, which has denied them the benefits of cultural contacts with other people.

It is important, therefore, before one attempts to make comparisons between nations, racial groups, and individual people, as to their cultural achievements, to examine closely the history of their experiences and the cultural contacts that were available to them. Franz Boas (1858–1942), one of the most distinguished and influential American anthropologists, who was a specialist in the culture and languages of American Indians, wrote in his work *Racial Purity*:

> The history of mankind proves that advances of culture depend upon the opportunities presented to a social group to learn from the experience of their neighbours. The discoveries of the group spread to others, and, the more varied the contacts, the greater are the opportunities to learn. The tribes with the simplest culture are on the whole those that have been isolated for very long periods and hence could not profit from the cultural achievements of their neighbours.

We are apt to take it for granted that our western culture is necessarily the most favourable for our species. This, however, remains an open question to which further reference will later be made.

On the Assets of Diversity and Equal Opportunity

THE MANY VARIETIES OF MANKIND enrich our species as a whole, for communities that differ in their genes are likely to have different aptitudes and talents. If, for example, everyone were gentically alike and had a great aptitude for mathematics to the exclusion of artistic ability, the world would have no painters nor musicians, and would be the poorer for it.

The great variety of genotypes to be found among the world's population not only places men into different environmental niches, but also enriches human experience. It is certain that some individuals of all racial groups can achieve high standards of education enabling them to be trained for most occupations. This, however, does not mean that persons cannot also have preferences and special abilities which are genetically conditioned. Most people, if trained from an early enough age, can, for instance, develop into athletes, yet only a few will excel and reach world championship standards. Many people can reach an adequate standard in physics if they receive proper instruction, yet only a few will become Nobel laureates, and one in a generation of men may become an Einstein. The boy who fails his examinations in mathematics may become a poet, while a brilliant mathematician may find the writing of poetry a difficult task. What is accomplished easily by some requires great effort from others, and it is these differences between people that are often genetically determined.

In the light of modern biology neither the racists, who maintain that cultural backwardness among peoples is genetically determined, nor the equalitarians, who carry the banner of identical ability, are right. There is for example, nothing unfair in providing gifted children with a different education from the average and treating average children differently from retarded ones.

What is important is that all children, regardless of race, social origin or economic status, should be given the opportunity of the best kind of education commensurate with their abilities.

Equality of opportunity among the peoples of the world is not a biological phenomenon, but an ethical and sociological ideal. To be equal does not necessarily mean the same thing as to be alike.

Different groups of people will always wish to arrange their lives in accordance with their diverse notions as to what form of happiness they wish to pursue. Their various contributions to mankind's store of knowledge will differ in kind, and the values placed on them will be relative to the observer's outlook.

Experiments on Built-in Preferences

PROFESSOR P. H. KLOPFER AND PETER KILHAM studied the reaction of yellow and black varieties of chicks with respect to the preference they showed for each other.

They found that chicks of either colour which had been kept in solitary confinement showed no consistent tendency to approach other chicks of the same variety, but if such chicks had been reared in daylight with chicks of their own variety they did develop a preference for their own kind. If, on the other hand, they were reared with chicks of the alien variety no consistent preferences appeared.

These experiments were carried out under carefully controlled conditions. They indicated that chicks seem to possess a built-in preference for their own kind which only became apparent after they had visual contact with their own kind. These preferences, therefore, only developed after a particular experience, in this case visual contact. When the results of these experiments became known some journalists queried whether such built-in preferences might apply to the human species, implying that human racial colour prejudices may be genetically influenced. This was a legitimate question, but it was unfortunate that many who read it were only too eager to find their own affirmative answers. To date, no evidence on this point exists, neither is it likely that we will be able to formulate experimental techniques to answer this question, for we cannot keep children in solitary confinement and observe their reactions when they are confronted with children of different colour.

All the same Kilham and Klopfer's study on black and white chicks has dispelled an old myth that if we could raise children in isolation from all stimuli, and then test their behaviour, we would find out what was truly genetic and what was not. Kilham's and Klopfer's study illustrates beautifully that, although a piece of information is transmitted genetically, it may still require an appropriate environment to manifest itself.

Nor is this the only example that organisms possess built-in preferences which require activation by external stimuli.

The Danish scientist Holger Poulsen, when studying the behaviour of birds, made the discovery that cock chaffinches which normally begin to sing in Denmark about the middle of February when raised in isolation attempted to sing by the middle of January, and two weeks later were producing an imperfect chaffinch song. In the middle of February he placed them together with singing linnets. Within a few days he noticed that they were imitating the linnet song with some success. But when he allowed them to hear an adult male chaffinch they immediately changed to the chaffinch song, nor did they ever again imitate the linnet. Experiments with other birds have shown similar results, which suggests that many of them possess a built-in preference for their own songs which requires a stimulus by a particular experience to bring it into the open. In the case of the chaffinches the experience takes the form of the chaffinch song. Evidence is continuously accumulating that most organisms possess built-in preferences which require activation by external stimuli.

Culture, Human Emotions and Race

A STEADILY GROWING NUMBER of scientists support the idea that our emotions, like our physical build, are an evolutionary adaptation of significant importance to our species. In this context I refer to emotions as our genetically determined capacity to form emotional bonds. This includes such attributes as the human capacity for love and affection for infants and children, affection and love between adults, and tolerance, a capacity to allow others to proceed with their lives without interference.

When animals of the same species do not fight with their own kind, they can be said to tolerate each other. Respect, being a specifically human quality, is also an enlightened form of tolerance. Man now takes these emotions for granted, associating them with civilization and his cultural achievements. Yet exactly the opposite is true, particularly for emotional bonds formed to aid in the protection and care of infants and children. These must have been strongest in primitive man, when such emotional bonds were still essential to his survival.

Because human cultures have enabled our species to substitute artificial aids for many of the emotional bonds without which non-human primates and Stone Age men could not have survived, genes associated with these emotional responses will be diluted during the further evolution of man and may even be lost altogether.

Emotional bonds formed between animals of the same species have long been observed by zoologists. Recently this subject has received increasing attention from students of animal behaviour, and studies concerned with monkeys and apes are both of interest and, as we shall see, of relevance to man. Let us investigate some of the findings.

All existing primates spend their lives in groups. These primate groups develop powerful adaptive mechanisms and emotional bonds which keep the groups together and have great selective advantage. Further, the formation of such bonds comes about easily and appears to give enjoyment to primates.

Drs S. L. Washburn and I. DeVore, who have studied baboons extensively in the wild, point out that the whole troop acts as a survival mechanism and its competence as a group far exceeds that of any individual. They also attach considerable importance to the protection of the entire group by the powerful adult male baboons. DeVore says:

> Once a monkey group begins living on the ground the much greater danger from predators would alone be sufficient to account for a more strict or organized social system, one which, for example, placed a premium upon male specialization in group defence.... While observing free-ranging baboons, Washburn found that the adult males were continually solicitous of the welfare of their young, especially when the troop was moving—a situation of more than usual danger. The females

> with infants, not necessarily those in estrus [breeding condition] stayed nearest the protective, dominant males.*

Dr Carpenter's observations of howler monkeys also showed the role of the adult males in keeping order inside the group, as well as protecting it from enemies. Of particular importance from the evolutionary point of view are the following observations of Washburn and DeVore concerned with the high incidence of injury and illness among living primates in the wild.

> When the troop moves out on the daily round, all members must move with it, or be deserted. We have seen sick and wounded animals making great efforts to keep up with the troop, and finally falling behind. At least three of these were killed, and the only protection for a baboon is to stay with the troop, no matter how injured or sick. In wild primates injuries are common and animals which are so sick that they can be spotted by a relatively distant human observer are frequent. For a wild primate, a fatal sickness is one which separates it from the troop.*

These and similar field observations on non-human primates indicate that natural selection must have operated very much in favour of individuals strongly motivated to keep together with their own kind. It also suggests that group-living has conferred powerful advantages upon monkeys and apes—the more highly developed primates. Among the most important of these are protection against predators, defence of territory against groups of the same species (this being closely linked with obtaining an adequate food and water supply), conditions enabling sick and injured individuals still capable of remaining with the group to recover in relative safety, ensurance of adequate reproduction, and the care and training of infants and juveniles. Doubtless there are many other advantages not as yet recognized.

Whilst we cannot automatically extend observations on non-human primates to man, primitive nomads, who have not yet reached the agricultural stage and still depend on a hunting and collecting economy, live in bands of 20 to 50 persons and establish rights over the territories within which they migrate. The selective advantages listed above largely apply to such groups of men and must have applied, even more strongly, to the many varieties of men of the late Pleistocene and before.

* *Primate Behavior* edited by Irven DeVore, published by Holt, Rinehart and Winston, Inc.

In Stone Age human societies, just as in present-day higher primate societies, those individuals who formed powerful attachments to others in the group were more likely to remain with the group and survive to pass on their genes to the next generation.

A baboon separated from his troop is not likely to survive for very long and this, too, must have applied to primitive man.

How can we translate the rather vague term 'emotional response' into a more precise concept? Let us for this purpose turn once again to observations of primates in the wild. When wild baboon troops are on the move the infant can only remain with the mother by clinging tightly to her fur and body. It has no alternative. Baboon mothers cannot hold their babies because they use their front limbs for locomotion. Thus any infants born with weak clinging responses will be quickly lost without an opportunity of passing their genes on to the next generation. Baboon infants, therefore, cling to survive; and in order to cling they are likely to form an attachment to an object that provides the opportunity for clinging. In the wild this is always the mother.

Dr Jane Van Lawick-Goodall, who for many years studied the behaviour of free living chimpanzees in the Gombe Stream Reserve in Tanzania, gives a graphic description of mother–infant relationships in these apes:

> Parturition (birth) has not been observed in the wild, but two mothers were seen with infants which had been born earlier the same day, during the night, or the previous evening. One of them, Sophie, was observed for some 10 minutes. She was a multiparous female and supported her infant efficiently, moving with the 'hunched gait' rounding her back and taking short steps with partly flexed legs so that the infant was supported by her thighs. The infant appeared to be clinging well with its hands. The second mother, Melissa, was under observation for 30 minutes, during which time she supported her infant continuously both with her thighs and by keeping one hand under its back and shoulders. She sat down at least once in every fifteen steps (I followed her for approximately half a mile). When sitting she cradled the baby closely, keeping her knees bent up, one arm or hand behind the infant's shoulders, and her feet close together under his rump. He was thus almost completely encircled by her body, legs, arms and feet. On the following morning Melissa's behaviour was similar. Once she momentarily removed her hand from the infant's back —he at once lost his grip with both hands and would have fallen had not the mother caught him.

> An infant, during its first three months or so, is almost invariably supported by the mother whilst she is sitting or lying . . . At times a mother may bend up one or other of her feet to give added support to the back or head of her infant. Whilst thus cradled, the infant may sit up straight on the mother's lap, its feet and hands holding hair on either side of her body between her armpits and groin . . . Whilst lying, mothers normally hold their infants close with one arm, or cradle them in their groins. Flo, when lying with Flint in her groin, invariably flexed one or both thighs and held one or both of her own feet with one hand in order to maintain the cradling position. Cradling behaviour enables the infant to relax. Thus although infants frequently retained strands of the mother's hair between their relaxed fingers while sleeping, they often ceased to hold at all . . . When the mother of a sleeping infant stands up, she often keeps her hand under its back for several yards until it grips tightly again.
>
> As the infant grows older, the mother gradually relaxes her cradling position and the infant itself must climb to a comfortable place on or beside its mother.

Dr J. Van Lawick-Goodall's observations also show that the mortality rate among infant and young chimpanzees is probably high in the wild. She observed chimpanzees during their infancy and up to the age of 4. Seven of these infants died during the first 4 years of their lives as the following table shows:

Age (in years)	*Number of known individuals*	*Number dying in that year*	*Mortality* %
0–1	13	4	31
1–2	11	1	10
2–3	7	1	14
3–4	8	1	12

Clearly this relatively high mortality among young chimpanzees makes for strong selection pressure against any mothers or infants who lack the capacity to form strong emotional bonds with each other.

If one extends these situations to early man, the importance of the formation of emotional bonds between mother and infant becomes even clearer. Since the human mother is practically hairless, a human infant has nothing to cling to; what is more, its arms and legs are ill-equipped to permit effective clinging. Human infants, therefore, are even less able to fend for themselves than monkey or chimp infants.

FIG. 90. Various supporting and cradling positions of infant chimpanzees under 3 months old. (After Dr J. Van Lawick-Goodall by permission of the publishers Baillière, Tindall & Cassell.)

Nevertheless, this problem had to be resolved in order that man should survive. If the infant cannot hold on to the mother, then the mother must hold the baby. To do so, the mother must feel a desire to hold the baby: she must derive enjoyment from doing so and experience frustration if she is deprived of this activity. At this stage it is important to remember that human babies are likely to have been as helpless as they are today for at least the last one hundred thousand years and probably even longer—in fact long before man's cultural achievements ever provided artificial aids for the protection and care of our young.

Man's evolutionary survival and that of many non-human primates has depended for thousands of years upon the successful evolution of emotional bonds designed to secure the safety and care of the newborn and the children.

The emotional bonds between human infants and their mothers must have had strong survival value in the past. It is quite legitimate to ask whether other human emotional responses—for example love and affection between adults, compassion for the sick and injured, devotion to a group—had similar survival value for our species. Since *Homo erectus* and the men that followed him depended on each other for survival in much the same way as today's nomadic hunters and non-human primate societies, the answer is likely to be yes.

If genetic studies appertaining to the emotional nature of man were to be undertaken, the results may well show that so-called primitive peoples of the world who lack all modern aids, and are still dependent for the care of their offspring on the emotional bonds which they acquired through natural selection, have retained a superior genetically determined capacity for feelings and emotional responses towards their offspring than in modern societies where the physical safety and well-being of infants and children has long ceased to depend on bonds formed between mothers and infants and is largely replaced by the mechanics of modern cultures.

As cultures advanced, so more and more human groups became less dependent on their emotional complement for the care of their offspring. Mothers who do not feel satisfaction when cradling their babies in their arms are not likely to do so, yet these infants will survive safely in our modern world.

As it becomes less important for us to possess the affection bonds without which Stone Age children could not have survived, genes associated with these bond-forming mechanisms will be diluted and may eventually be lost.

FIG. 91. Transport of infant chimpanzees. (a) and (b) Infant about 5 months old clinging to mother's arm and dangling as she walks. (c) Mother carrying infant 7 months old who has refused to cling on. (d), (e) and (f) As she moves off mother gestures to infant 1 year old to climb on her back. (g), (h) and (i) Infant climbing on mother's back. After Dr J. Van Lawick-Goodall, by permission of the publishers Baillière, Tindall & Cassell.

The emotional make-up of man will be as important to our survival as a species as our ability to solve problems. Paradoxically this will be especially so because our technology has enabled us to develop ultimate weapons of destruction which will require to be carefully controlled. There is more chance of achieving effective controls if human populations are capable of those and similar strong positive emotions that helped our ancestors to survive.

If studies should eventually show that men vary genetically in their emotional capacities, it will be the primitive societies (most of which are coloured and have often been classed as backward) who are likely to be better endowed with gene complexes that make men capable of forming close bonds with each other. These societies are still largely dependent on their emotional make-up for their survival, something which is no longer true in societies with highly developed cultures.

Apart from our intellect it is our capacity to form close and permanent relationships with each other that makes us truly human—a fact worth remembering if scientists should ever establish significant intelligence differences between human races.

Is man more human now because of his cultural excellence or because of certain emotional attributes which fall under the broad heading of bond-forming capacities between individuals and groups? This is a question of ethics and not biology. But if a choice had to be made between men of high problem-solving ability capable of the most sophisticated technology but lacking in the ability to feel certain bond-forming emotions, and those of low problem-solving ability but endowed with genes allowing for the full development of emotional qualities which enabled primitive man to survive, I would choose the latter. Such men would be less likely to destroy one another than the unemotional technocrats. Ideally, men should possess gentic attributes enabling them to develop into emotionally balanced and intelligent individuals, and of course an educational and social system which would develop both to the maximum extent.

The question remains whether those human groups where intelligence has led to sophisticated cultures may not be in danger of gradually losing the gene complexes which favour those emotional ties to which civilization has always attached great ethical values.

Race Discrimination

THE TERM 'RACE', particularly in our Western society, has now become heavily tainted with emotional overtones, and children are conditioned at an early age to believe in the existence of race differences. Even such an innocent nursery rhyme as 'Eena, meena, mina, mo, Catch a nigger by his toe' implies racial inequality and can hardly fail to escape the attention of children. Sadly, little or nothing at all is taught in most schools concerning the facts relating to the varieties of man. While it is important to learn where the major oilfields of the world are situated, or which nations are the largest steel producers, it is even more important to teach children about man's varied cultures, and how these are valued by human societies. Our aim also should be that children are given the opportunity to learn those facts about races that have been adequately established by scientists. Prejudices early acquired are notoriously difficult to eradicate. Children must be taught that a certain kind of nose or skin colour does not make one man either better or worse than any other.

Physical differences are often the nails upon which culturally generated hostilities are made to hang, ending with the smug and empty conviction that a superior race is one that looks like oneself, and an inferior race is one which does not.

While we may admire the Americans for their advanced technology which allows them to send men to the moon, we should be able to show equal respect and understanding for still primitive communities of men such as the Australian Aborigines or African Bushmen who have not attained sophisticated scientific cultures, but have thought to enrich their lives by rather simpler activities which may give them equal satisfaction and perhaps less coronary thrombosis and fewer duodenal ulcers.

To be human means to understand, and to understand should mean to respect. A basis for a solution to the racial problems of the world lies in recognizing the importance mutual respect plays in relationships between the various human types whatever their cultural achievements. Dr Ashley Montagu makes these comments:

> The plea for fairness in dealing with ethnic groups not our own is usually phrased in terms of 'tolerance'. But if we are to make progress in ethnic relations it is desirable to recognize that tolerance is not good enough, for tolerance defines an attitude which constitutes a somewhat

reluctant admission of the necessity of enduring that which we must bear, the presence of those whom we do not like. A New York high school girl puts the matter in a nutshell. 'Tolerance', she said, 'is when you put up with certain people but you don't like to have them around anyhow.' . . . Tolerance is the attitude of mind of those who consider themselves not only different but superior. It implies an attitude toward different ethnic or minority groups not of understanding, not of acceptance, not of recognition of human equality but of recognition of differences which one must suffer—generally, not too gladly.

Therefore to be tolerant of other people does not seem to be enough, one should at least attempt to be fair, and what is more important still, one should try and respect where such respect is due irrespective of race or social background.

Race discrimination also has economic causes and is often aggravated by economic competition. But I believe that much superficial prejudice is closely linked to differences in human outward appearance, and ignorance as to the nature and meaning of this.

Racists usually hold their views because of early conditioning, and this includes education. It is their social environment that makes men think in the way they do, and their discriminatory attitudes towards other races are usually based on biological misconceptions about races in general, which are acquired by hearsay.

Peoples all over the world pursue racial policies of some kind. Prominent during this century has been the German persecution of Jews, although these do not at all represent a racial group in the biological sense. They represent an example of a group of people who by virtue of their religious beliefs have for hundreds of years intermarried. Thus they have exchanged their genes among themselves and evolved more or less as a unit. This explains why some Jews may have certain physical characteristics in common.

Most prominent of all race discrimination is that of white men against Negroes and other coloured peoples. Recently, much has also been heard of African discrimination against Asian minorities in some of the newly formed African states.

As mentioned earlier, the motives for racial discrimination vary widely and include many shades of social and economic causes, but clearly the most soul destroying of all racist doctrines is that which labels a people inferior because of an allegedly inferior genetic make-up.

It is one thing to belong to an Asian minority in an African state and find oneself discriminated against because of economic success; it is yet another to be a Negro in a white dominated society and be told that you are inferior because of your allegedly inferior genes or merely because of the colour of your skin.

All race discrimination is evil, but that which springs from a conviction of genetic superiority is especially so, for it deprives the victim of self respect. It has been largely practised by white men against Negroes and other coloured people.

Arguments for and against the Use of the Term Race

THE TERM 'RACE' is today heavily tainted with overtones of injustice and discrimination, largely because it has popularly been used to describe minorities which on occasion, through no fault of their own, have attracted the displeasure of the majorities.

Because it is difficult to find a precise meaning for the word 'race' many people feel it is best to discard the word altogether and to pretend that mankind has no races at all. Others hold the view that because the word 'race' has been so badly misused by various hate-mongers, and has become associated in the public mind with unscientific notions and prejudices, its use should be discontinued.

Dr Ashley Montagu strongly champions the idea that the words 'ethnic group' should be substituted for race. He defines ethnic group as one of a number of populations which together comprise the species *Homo sapiens* but individually maintain their physical and cultural differences by means of isolating mechanisms such as geographical and social barriers.

Dr Montagu feels that one of the most important advantages of the term ethnic group is that it eliminates all question-begging emphasis on physical factors or differences, and leaves that question completely open while the emphasis is now shifted to the fact—though it is not restricted to it—that man is predominantly a cultural creature.

There is much validity in the argument that it will always be difficult to remove the emotional overtones which have been associated with the term 'race'. It would be equally difficult to remove the horror and fear now associated with the drug thalidomide which has been responsible for the tragic malformations in infants, even if it were discovered that its use could be beneficial to man. If ever such a situation arose with any drug it is likely that the manufacturers would remarket it under a new name.

So it may also be with the term 'race'. If in future we call races ethnic groups, it may help to remove some of the emotional overtones associated with the concept. These attitudes therefore are not unreasonable, but it is doubtful whether a change in name will be sufficient to solve the problem of race prejudice or if the change in name would be accepted.

Especially because the word 'race' has now become associated with unscientific notions and prejudices in the public mind, it would be better to attempt to set the records straight rather than to escape from them altogether by substituting a new word with a new meaning. All mankind belongs to one species which, like many other animal species, has through the course of evolution become partly divided into a number of different types. These have been called races. They are less rigidly integrated or defined than species, but they remain nevertheless distinguishable populations which form loose breeding communities within which matings occur more frequently than between them. They do not represent precise biological phenomena but they nevertheless exist. A more constructive approach would be to attempt to educate people in such a manner that eventually they could come to accept racial differences, whether external or internal, as part of our human heritage, and would look upon the large variety of men that make up our species as a diversity to be proud and not ashamed of.

On the Need and Desirability to Study Race Differences

THE MAIN DIFFICULTIES scientists face in comparative racial studies between men lies in our ignorance of man's genetic make-up. To date, scientists have been unable to formulate a precise definition for any group of men in terms of their genes, and thus they lack a scientifically valid basis for such studies. While it can be shown how blood group genes are passed on from generation to generation, there are only a few other human genes that can be traced in this way. If we could identify genes that made for intelligence, and understood how these were inherited, matters might be different.

Because at present our methods are inaccurate, and results of comparative intelligence studies cannot provide us with the scientifically valid answers we seek, some sociologists regard these studies as a dangerous social evil because their psychological effects on human populations could be disastrous.

In a report published by the National Center for Educational Statistics in the U.S.A., Professor James S. Coleman said:

> The psychological reactions of a Negro in American society to the fact of his being a Negro occur, not as a result of the particular set of genes he possesses, but as a result of the fact that he is identified by others as a Negro, comes to identify himself as a Negro, is reacted to by others as a Negro and believes himself a Negro. To this, I think, we should add that any studies which proclaim to show Negroes to be of inferior intelligence, and which clearly cannot be based on sound genetic evidence because to date this simply does not exist, will not make a Negro feel inferior because of his social position but also because he may regard himself because of his black skin as of inferior intelligence.

At a recent symposium in the United States on 'Science and the Concept of Race' the importance of basing comparative race studies on sound genetic identifications of human groups was repeatedly stressed and the difficulties facing scientists in this connection were high-lighted. Professor M. H. Fried expressed his convictions that a method of racial typing based on externally visible features such as different colouration, hair form and colour, eye colour, eyelid formation, nasal structure and size, etc., is totally inadequate, and does not form a sound basis on which further studies as to the intelligence of such different groups can be based.

He quotes an example of studies of this kind carried out in Brazil, where it was found that if one used these external distinguishing marks to differentiate between racial categories one was liable to end up with several hundred different types. What was even more confusing, different students were liable to place the same individuals into different categories.

In relation to the American Negro, Professor Fried principally objects to the methods of those workers who have categorized the American Negro as a recognizable and clearly defined group, and have set up the criterion of membership in this group by the definition of 'more-or-less' African ancestry.

Fried's main hope in relation to such work lies in a young generation of scientists concerned with studies of new physical anthropology which includes a number of different disciplines such as physiology, ethnology and genetics. These young scientists are thoroughly proficient in their subjects and are likely to produce answers to some of these questions which will apply to small populations well controlled for genetic information. Dr Fried believes that this new generation of scientists should begin to meet the criteria for truly scientific studies of populations. But he points out that these will be studies of populations, and not races.

The dangers inherent in poorly executed race studies are also stressed by the distinguished geneticist Professor T. Dobzhansky. Nevertheless he feels that failure to use modern methods for scientific race studies is not a reasonable policy. In his summing up address at the symposium he said: 'If scientists refuse to study races we shall have pseudo-scientific studies of races by race bigots.'

Professor Dobzhansky also stressed the point that there is abundant evidence to show that people exist within each human group who possess potentialities for cultural and intellectual development equal to or greater than the average potentialities for any other group.

In her concluding remarks at the end of the symposium Dr Margaret Mead, the well-known anthropologist, referred to the question whether information should be withheld by scientists if it may have bad social implications. She said this:

> I believe that information should not be withheld, but that does not absolve the scientist of responsibility for what misuses this information is subjected to. In other words, if I publish certain information, and I see that others are making out of this information what I regard as unjustified use, I would consider it my obligation to say that that particular use, I regard as misuse. I would say that it is not justified to say, 'I am a scientist, I have produced information, and I am no longer interested in what happens after that.' That sort of absolute impartiality I believe amounts to absolute irresponsibility, and in our world a scientist has no right to be irresponsible.

Dr Mead concluded with these significant comments:

> The price in human suffering, loss of human potential and ethical damage that this country is paying because of such concepts as 'average biological differences' applied to parts of our highly mixed and diversified populations, is incalculable. A fuller understanding of the significance of the history of our unique species, our vicissitudes in time and

place, and our inherent potentialities is an essential underpinning for dealing responsibly with our relationships one with another.

She was speaking about American society but these comments, so wisely made, surely apply to all peoples of the world where racial injustices are perpetuated for whatever reasons.

6

Culture and the Present State of Human Evolution

THE RAW MATERIALS operative in natural selection are provided by mutations and sexual recombinations of the hereditary units, the genes. Mutations, sexual recombinations and natural selection were the principal factors in the evolution of modern man and all other living species. Already some of the hominids which preceded him were capable of making and using tools and must have acquired some means of cultural transmission. The next evolutionary step, however, marked a turning point in his fortunes. A new type of man now appeared on the scene, capable of creating a diverse technology and numerous forms of symbolic communication, enabling him to create a sophisticated culture. Other forms of life adapt to their environment by taking advantage of gene changes in accordance with the requirements of their natural surroundings. Only man, through the evolution of his brain and his intelligence, can also adapt by altering his environment to suit his genes. He can, therefore, also create his own environment. It took him a long time to reach this position, however, and the process started very gradually.

At least 1,700,000 years ago *Australopithecus* used crude bones as primitive tools. The various forms of *Homo erectus* that followed him became accomplished tool-makers. Their still-primitive hand axes and flake tools have been discovered throughout vast areas of Africa, Western Europe and Asia. The makers of these artifacts were hunters and food-gatherers. About seventy-five thousand years ago, during the time of the

Neanderthal men, man's tools became sufficiently specialized to suggest that they were adapted to food-getting needs in varying environments.

The emergence of our sub-species *Homo sapiens* goes hand in hand with greatly improved blade-like tools incorporating a high degree of usefulness and skill in their manufacture. Using such new instruments these men must have embarked upon more systematic food-gathering methods and organized hunting of animals than did their simple, food-scavenging predecessors. As time passed human populations became increasingly accomplished in their task of cultural invention and adaptation, and this enabled them to adjust to environments as different as the tropical deserts and the Arctic tundra. No doubt at this stage natural selection was still responsible for most of the adjusting. Yet, even in those early days, man attempted to adapt his environment to fit his genes by building shelters and using fire to enable him to survive adverse climatic conditions.

The successful adaptation of our species to different environments resulted in still greater cultural complexity and diversification, and between eleven and nine thousand years ago some communities in the Middle East began the first experiments with domesticated plants and animals. They had finally reached the threshold of food production—a prerequisite to settled village life.

At first, the impact on the environment by our species was slight. As populations increased in size, however, so the impact of human works on the environments grew stronger, until today there is hardly a corner of the world where men's handiwork is not in evidence.

The increase in human population during the last 25,000 years has been truly staggering, as the following figures demonstrate.

25,000 years ago	3,500,000 (estimated)
10,000 ,, ,,	5,500,000 ,,
6,000 ,, ,,	86,000,000 ,,
2,000 ,, ,,	133,000,000 ,,

From then onwards the increase becomes almost frightening.

300 years ago	545,000,000 (estimated)
200 ,, ,,	728,000,000 ,,

By 1800 the figure had reached 906 million. At the beginning of this century world population was estimated at 1,610 million. By 1950 the figure had soared to 2,400 million. Today we have passed the 3,000 million mark, and it has been variously estimated that by A.D. 2000, our species will have touched the staggering world population figure of 6,000 million—unless before that time the majority of the world population will adopt methods of birth control.

The phenomenal increase in our numbers, particularly in the last three hundred years, is due largely to the advance of culture creating conditions favourable to the support of such multitudes of our species. Yet, like all other living organisms, man remains the product of his biological inheritance. The most important feature of his present evolution is that his genes continue to mutate as they have done throughout his evolution. It is known, as mentioned earlier, that certain special agents can induce mutations. Ionizing radiation and chemicals such as mustard gas have been mentioned, and more man-made agents are constantly being discovered. As a direct result of this, populations living in industrialized areas may be particularly subject to increased genetic variability through rising mutation rates. At present, because of past wide-scale nuclear-bomb tests, practically the whole world population is variably affected by ionizing radiation from radioactive fallout, which includes such damaging substances as strontium 90. This is readily absorbed into our bones because the organism cannot distinguish the element strontium from calcium, one of the chief constituents of our skeleton.

Apart from mutations induced by our man-made environment, some mutations occur spontaneously in people not exposed to any special agents. Studies have revealed that approximately one gamete in every fifty thousand produced by a normal person carries a new mutant gene causing retinoblastoma—a cancer of the eye which affects children. Many other mutations are known which produce hereditary diseases or malformations, while others are likely to induce various kinds of constitutional weakness. A few, however, must be useful, otherwise evolution could not have occurred. Indeed, such useful mutations have been demonstrated and observed in experiments with organisms such as flies and bacteria (see Chapter 3). There are also those mutations which produce minor variations in the human form such as baldness, lighter or darker skin colour, differing blood types and others. Such traits may at first seem neither useful nor harmful, but caution is required when evaluating these seemingly neutral traits. For example, Sheppard and Clark of Liverpool University found evidence that people with blood type O may suffer slightly more from duodenal ulcers than the general population. It does not follow, of course, that blood O is necessarily bad. It is, in fact, the most frequent blood group found in many populations and may even confer certain advantages yet to be discovered. Also outward mutations may be associated with others which are not immediately appreciated but nevertheless of value.

Some mutants can be highly dangerous when present in the homozygous condition where the same type of gene is inherited from both parents. In the heterozygous state, however, when only one parent carries the gene, it confers considerable advantages upon the offspring. The sickle cell gene discovered by Antony C. Allison of Oxford University is such an example. This gene when present in the homozygous state causes the fatal disease, sickle cell anaemia, yet in the heterozygous condition it renders the individual reasonably resistant to certain forms of malarial infection. Found very frequently in the native inhabitants of the Central African lowlands, where malaria has long been endemic, it is relatively rare in the populations of the more healthy highland areas. It is certain that other such genes exist in human populations which can play a dual role, depending on their heterozygous or homozygous state; but as yet very little is known about these.

Biologists are generally agreed that any increase of random mutation rates, no matter how small, can only add to human disabilities due to defective heredity. This is the most important reason for condemning ionizing radiation from nuclear fallout, and why the subject has recently attracted worldwide attention. X-rays, too, have been carefully scrutinized for their effects. Clearly, however, mutations caused by human actions form only a part of the sum total of mutations, some of which as we have seen arise 'naturally'.

The process of evolution is based on the countless mutating genes that have arisen since life on earth began. Mostly those useful to a particular organism have been preserved, and these represent but a minute fraction of the total that have taken place. Clearly mutations that affect an individual so adversely that he can no longer survive in his environment are quickly discarded by the extinction of those which carry them. The principal factor that preserved useful mutations and eliminated harmful ones was natural selection. A question which requires further examination is whether natural selection is still operating in mankind today to contribute towards its fitness to survive in a given environment. Mutations of little value when first established may sometimes be 'pre-adaptive', that is of value in a future environment.

In this context it is as well to examine the meaning of the word 'natural' in relation to the environment. Colloquially, it is often used to describe environments untouched by human hands. In our sense, however, it refers to all environments of a species, including those made by man.

Also the notions that it may be more natural (and therefore healthy) for man to toughen his constitution by discarding warm clothing in cold weather or to eat raw foods in place of sophisticated cooked meals can be very misleading.

For our, and other, species 'natural' is that which will allow reproduction to proceed most successfully. If this should mean that in a thousand years from now we might find ourselves in glass domed enclosures on another planet or satellite, where living conditions have proved more successful, and our species can perpetuate itself more satisfactorily by feeding entirely on man-made proteins, then this new home and diet will be our 'natural environment'. In the last twenty thousand years man's environment has altered enormously, mostly by his own making, and it would no longer be natural for us to try and live like the Cro-Magnon men of the Upper Palaeolithic.

Clarification should also be sought for such popular phrases as 'the struggle for life' and 'the survival of the fittest'—words applied by Darwin to animals which metaphorically might struggle against a freezing climate and were selected by evolution's experiments for the genes that enabled them to grow warm fur. The Stone Age men, too, struggled 'for survival'. Here selection pressure favoured genes that were making for keener intelligence. It was the so-called social Darwinists—Darwin was not one of them—who applied the concept of the struggle for life and the survival of the fittest to man-made violence and warfare—an idea which has long been discredited.

It is true of all higher forms of life, and particularly of man, that the so-called struggle for existence can be won not only by competition and strife but also by mutual assistance. This concept, if equally applied to man in his Stone Age culture and today, leads to the conclusion that although a physical contest between two individuals is likely to end in favour of the physically stronger and more aggressive, the society of the loser may well prove to be the more successful in the long run. Especially is this so where a tradition has grown up that neighbours will offer mutual assistance to each other in their day-to-day problems. The hunters of the Palaeolithic depended on such co-operation to kill their often fast and powerful game.

It is also noteworthy that the modern interpretation of Darwinian fitness, or adaptive or selective value, refers to relative reproductive capability. The following example illustrates that Darwinian fitness should be understood to mean reproductive fitness.

The hereditary malformation known as achondroplastic dwarfism is caused by a gene mutation that produces individuals with normal heads and trunks but short arms and legs. Adults who suffer from this hereditary debility may, however, otherwise be completely healthy. Yet Dr E. T. Mørch, a Danish research scientist, discovered that achondroplastic dwarfs produce on the average only some 20 surviving children as against every 100 children produced by their normal brothers and sisters. Exactly why this state of affairs exists is only partly understood. Clearly this is very strong selection, and is important from an evolutionary point of view because persons afflicted with this disability are much less efficient in passing on their genes to succeeding generations than are unaffected individuals.

Genetically speaking, therefore, the surviving fittest need not have a genetic make-up which renders him some sort of superman; he is merely the parent of the largest surviving progeny. These definitions are important if we are to consider the question of whether natural selection is still active in human populations and how such selection might operate.

Let us now consider two artificial situations applicable to man which could preclude natural selection. If all adults married, and each couple produced equal numbers of children, all of whom survived and married, and in turn again had the same number of children, and so on, selection would be eliminated. In another situation artificial selection would replace natural selection if the number of offspring were to be dictated to each person either by himself or some outside authority on the basis of the usefulness of his hereditary endowment. Viewed on a global scale this type of artificial selection applied to the human species seems a very long way off (it is of course continuously applied in animal husbandry) and in the meantime natural selection is proceeding. All types of natural selection proceed, however, within the context of environment. As the environment changes the Darwinian fitness of various heritable characteristics changes with it.

A fundamental difference between ourselves and the rest of the living world is that man by his own efforts can, and does, continually change his environment, thus altering the selective pressures for and against certain of his genes.

Similarly, he is also changing the environments of other living forms, sometimes accidentally, more often by direct planned control. In this way he alters the selective pressure for and against certain genes of an increasing number of organisms.

One of the clearest and best examples of a form of culture which directly affects the evolution of man is the progress of medicine and public health during the last 100 years. Retinoblastoma, the eye cancer of children, was almost always a fatal disease until a treatment for it was successfully developed: since it is mostly a hereditary disease, natural selection in the past disposed of virtually all the harmful mutant genes before they could be passed on to succeeding generations. Today, with the advance of treatment methods, 70 per cent or more of the carriers of the retinoblastoma gene are able to survive and, if such persons marry and have children, they will transmit the defective gene to half their offspring. Two other examples illustrating the effect of civilization on our genetic make-up are afforded by progress in the control of tuberculosis and malaria. In the last century, tuberculosis, particularly in industrially advanced countries, was a terrible scourge. Some 500 people in every 100,000 died from the disease. A general improvement in living standards and housing conditions, and the discovery of the antibiotic drugs have sharply reduced the death rate of the disease in most advanced industrial societies to under 10 individuals per 100,000. Similar progress has been made in reducing the mortality from malaria which at one time afflicted a seventh of the world population. Both tuberculosis and malaria are infectious diseases, and thus are a hazard of the environment. Further, there is sound evidence that individual susceptibility, both as to contracting the infection and as to the severity of the disease, is to some extent determined by the genetic make-up of the individual. As earlier mentioned malaria is a disease the susceptibility to which is dependent on the presence or absence of the sickle cell gene. Clearly both in tuberculosis and malaria, the Darwinian fitness of susceptible individuals increases as the prevalence of the disease decreases.

Having earlier discussed the effect of culture on the mutation rates of our species, the fact emerged that many of our activities lead to an increased mutation rate and, since most mutations are harmful, to an ever-increasing number of harmful genes. Another effect of culture, particularly the advance of medicine and hygiene, is to decrease the rate of discrimination and elimination of harmful genes from the human population by the process of natural selection. Thus, culture seems to be responsible for a worsening of our genetic make-up as well as frustrating natural selection to rid man of harmful genes. Professor T. Dobzhansky, the geneticist, views the situation with apprehension, pointing out at the same time that the elucidation of the human genetic make-up and its possible deterioration poses many as yet little understood problems.

Matters become even more complicated and uncertain when we come to consider genes that are harmful in the homozygous condition but beneficial in the heterozygous state. The sickle cell gene is an example. Little is known today how many other hereditary diseases and malformations are maintained by the advantages their genes confer in heterozygous carriers.

Similarly, it is difficult to foresee the genetic effect of relaxing selection pressure as a result of human cultures. Thus, for example, tuberculosis is today largely a curable disease, but if susceptibility to tuberculosis is maintained among populations by recurring mutations, then our ability to cure this disease will probably increase the concentration of mutant genes that make for susceptibility. When certain genes bestow resistance on an individual if present in a single dose, but make for susceptibility in a double dose, the fate of such genes in populations might then be determined by unknown selective forces.

Clearly, whatever form our culture may take in the future it is certain that human genetic patterns will continue to alter under the shelter of man-made environments and, in particular, modern medicine.

Unless it will in time become possible by advances in molecular biology to actually eliminate harmful genes from germ cells, and in this way prevent the inheritance of diseases and malformations, it may well become necessary to practise some form of artificial selection. Whichever way man will eventually choose to solve his genetic problems, there is little doubt that his survival as a species will, in the future, increasingly depend upon his intelligence and the technology which springs therefrom. Whether man's course which leads him to ever more sophisticated cultures will ultimately be good or bad for his species, nobody can tell: suffice it to say that it is now much too late to do anything to reverse this situation, for we seem to be irrevocably committed to continue on a course of cultural advance.

What, if anything, then, can we do immediately about the increasing number of harmful genes in human populations? One thing is certain—we cannot deliberately allow children to die from a hereditary disease, such as retinoblastoma. Any society which, on eugenic grounds, would contemplate refusing a cure to those persons who suffer from any hereditary disease would be so devoid of compassion and ethical codes that they would lose the right to be called human.

A very different approach to the same question may be found through the processes of law whereby persons who are known to be carriers of harmful genes, such as those that may cause death to an individual unless he receives treatment, are prevented from having children. Such a step may be difficult to take, since it clearly encroaches upon the rights of the individual for the benefit of future generations. Whether societies will ever come to terms with ideas and laws designed to improve the genotypes of future generations will depend on the evidence which biologists and geneticists can submit for serious consideration. As a result of such evidence people should be able to decide by democratic means whether to apply legal sanctions.

It is noteworthy that many people who suffer from hereditary diseases and are advised by their doctors to refrain from having children do so voluntarily.

One remedy open to us now which would not offend ethics and which could help towards improving the situation, lies in an attempt to reduce the artificial increase in mutation rates caused by our cultures. The most obvious example that comes to mind refers to the partial ban on nuclear bomb tests above ground, now agreed to by most of the nuclear powers. Let us hope it will soon become a total ban.

Even more important is the need to make certain that the nuclear weapons that have been developed will never be used in war, for apart from the devastation they would cause, the nuclear fallout resulting from an indiscriminate use of such weapons would be highly dangerous from the genetic point of view, not only to man but also to the vast number of other living species on our planet.

The genetic basis of our culture is, in the end, a result of the intelligence acquired by our species through natural selection. This at present seems to bestow upon us considerable advantages over other forms of life. Yet natural selection, although a remarkable mechanism for adapting creatures to their environments, does not by itself guarantee their survival. Many more species have become extinct during the millions of years that life has existed than are living today. This has occurred without the softening influence of civilization: their downfall often being caused by over-specialization to suit a certain environment in which they may have lived comfortably and successfully for tens of thousands of years. Only when the environment changed did they find they could not adapt sufficiently quickly to meet the changed conditions. Thus they perished.

Our greatest hope lies in man's phenomenal ability to adapt himself to changing environments and his growing power to alter his environments to suit his needs. Our greatest danger lies in thoughtlessly creating an environment to which we can no longer adapt ourselves. Our intellect has largely been able to replace the blind force of natural selection, but our ideas as to what type of environment may be best for our species remain frighteningly vague. Imagine a situation where, as a result of a nuclear war, the earth became so polluted with radiation that in spite of man's tremendous technical knowledge insufficient time would be left to him to adapt biologically to a radiation-polluted environment. Without having to enter the realms of science fiction, this sort of situation might well spell doom for our species.

The dangers of excessive radiation seem obvious, yet there are many other cultural hazards not directly connected with gene mutations, such as the pollution of our environment with insecticides and other man-made chemicals which will pose problems. Perhaps, however, one of the greatest hazards to which the nations of the world appear to remain largely oblivious is our phenomenal increase in numbers which threatens to swamp the earth with human life within the next hundred years. Either man will regulate his own population density by voluntary means, or famine, pestilence and war will do it for him. Some means of limiting the world population will have to be found! Whether a solution to the world population problem will be forced upon us by major disasters, or whether we will use our intellect to voluntarily limit our families before such disasters occur, ought to depend on a world-wide awareness of the urgency of the problem.

In the end it will be man's actions that will determine his destiny. Clearly our biological evolution continues, but to predict where it is taking us would, to say the least, be rash. It will certainly depend far less on the processes of natural selection, known to us among organisms such as the peppered moth which adapted itself to sooty urban environments by acquiring a matching shade, or bacteria which become resistant to penicillin. Rather will our further evolution be increasingly dominated by our behaviour, which now depends largely on man's free will. This means that although our capacity to 'behave' is determined by our genes, the manner in which we behave is not in every instance gene determined. Attempts to forecast the effects of our fast advancing technology on the social well-being and happiness of the peoples of the world have been none too successful; how much more difficult will it be to forecast the effects of our culture on our invisible genes of which most of us are quite unaware.

Will our hunger for knowledge, which drives us relentlessly on in search of ever more sophisticated cultures, ranging over all fields of human endeavour, pay sufficient attention to the more fundamental problems concerning the effects of man's cultures on his biological evolution? Time will tell, but when all is said and done it will be positive solutions to these problems which will help to determine whether our species is really the success it appears to be.

Symbolic thought and self-awareness came to man slowly and resulted from the development of his brain over the last million or more years. Thirty-five to forty thousand years ago man appeared to have reached a stage when he became master over his own actions. This led to culture as we know it today, and made us the only species on earth able to contemplate the past and the future. It also saddled us with the tremendous responsibility of looking after our biological evolution and that of the many other forms of life.

Our responsibilities to other forms of life have been admirably expressed by thirteen-year-old Karen Lynn Craighead, the daughter of Dr John Craighead, who was studying the behaviour of the Grizzly Bear in North America to find ways of preserving the species. She simply said: 'We want to preserve the Grizzly because when he has gone we won't be able to make another one.' Put in this way the extinction of animal species has a horrifying finality, and our responsibility for their preservation is brought home to us with a clarity rarely achieved by all the literature that has been published on the subject.

The behaviour of monkeys and apes is complicated enough as has been so amply illustrated by the studies of C. R. Carpenter, Irven DeVore, S. L. Washburn, G. Schaller, Jane Van Lawick-Goodall and others. How much more complex is the behaviour of man which no longer largely depends on innate responses, but is to a great extent directed by his thought processes? If we contemplate those actions of man that lead to further developments of our culture, it becomes clear that they are based mostly on experimental processes of trial and error. In that part of culture called science this is immediately obvious, for most forms of scientific research are based on carefully planned experimental methods.

For example, early this century Paul Ehrlich (1854–1915), the German medical research pioneer, discovered the synthetic arsenical compound arsphenamine, known commercially as Salvarsan, which proved effective in rabbits and apes experimentally infected with the organism of syphilis. This at the time revolutionized the treatment of the human disease, although it was later superseded by the discovery of penicillin, which by experiment was shown to be even more effective if used in conjunction with other drugs.

Not so obvious are the experimental methods we employ in our personal relationships with each other. Yet here, too, a good many of our actions are prompted by a sense of the unknown and a desire to invoke further discoveries.

National governments experiment with new methods of administration, often unwittingly extending their experiments to the patience of the electorate. Rival powers test each other's determination by provocative acts. Even children will provoke in order to experiment upon how long their parents' patience will last. Many of our experiments are positive, seeking improved conditions for mankind. Others, like the conquest of space, are of debatable value. Some are negative, leading to anti-social results. Nuclear bomb tests are an obvious example.

Most of our actions are directed towards the evolution of cultures. Whether they apply to scientific research, the arts, national government, production methods on the shop floor or personal relationships, they contain an experimental element generally designed to exchange existing methods for more successful ones. The foregoing examples of culture are, however, extremely limited in their scope, because culture as such embraces a much wider field of our behaviour. It includes the sum total of habits, customs, beliefs, language and techniques used by man in all fields of activity and it is passed on from generation to generation as a result of having been taught.

Since our cultural experiments are now inevitably tied to our biological evolution, the latter also continues to be experimental. Now, however, most of these experiments stem from man's thought processes and are no longer at the mercy of nature's blind games of chance.

The difficulty mankind now faces relates mostly to our ignorance of which types of culture may be the most favourable ones for our species. Unless an advanced computer technology will, in future, be able to unravel such complex questions, and man is prepared to follow such computer advice, he will, as he has done in the past, have to leave the development of culture in the lap of chance in the hope that he will interpret the results of his cultural experiments to his advantage.

The German philosopher Friedrich Nietzsche once said: 'Man is a rope stretched between the animal and the superman—a rope over an abyss.' In the light of modern developments in science, technology and politics, Nietzsche's idea is remarkably topical. It remains to be seen whether the advances in these will eventually enable man to achieve the creation of an utopian superman whose sole purpose will be to create a culture wherein all men will be able to live in eternal happiness. A formidable task indeed and one from which I believe we are today further away than the Cro-Magnon men were from the development of a rocket ship that would take them to the moon. Indeed it may well be more important for us to concentrate on the Nietzschean 'rope' rather than on the superman goal on the far side of the abyss, for there is no guarantee that our passage will be accomplished in safety.

Glossary

ADAPTATION. A process by which an organism changes in structure and behaviour to fit to its particular habitat and environment.

ADAPTIVE RADIATION. Marked evolutionary changes in structure, physiology and behaviour in a single ancestral group of organisms to fit them into numerous distinctive and greatly diverse ecological niches.

ALLELE. One of two or more alternative states of a gene occupying a particular position on a chromosome.

AMINO ACID. An organic compound containing both an acidic carboxyl group (COOH) and basic amino groups (NH_2). Amino acids are fundamental constituents of all living matter because some hundreds or thousands of amino acid molecules are combined to make each protein molecule. Twenty different amino acids are common constituents of proteins and about twenty-five different amino acids are known.

ANATOMY. The study of the physical structure of organisms.

ANGULAR MOMENTUM. Momentum resulting from rotation.

Anthropoidea. A sub-order of the order primates including the monkeys, apes and man.

ARBOREAL. Living in trees.

Archosauria. One of six sub-classes of reptiles which includes the crocodiles and their relatives and many extinct forms such as dinosaurs, flying reptiles etc.

ARTIFACTS. A term used for tools uncovered in archaeological excavations.

ASTRONOMICAL SPECTROSCOPY. A major branch of astrophysics in which light from celestial objects is dispersed into spectra for the purpose of obtaining information concerning the sun, planets, stars, galaxies, comets and meteorites.

ATOM. The smallest part of an element that retains its physical and chemical properties.

Australopithecus. A genus of African fossil primates classed with the australopithecines that has some human and some ape-like features.

AUTOSOME. One of the chromosomes other than the sex chromosomes.

BIOCHEMISTRY. The study of the chemistry of living organisms and their processes.

BODE'S LAW. In 1772 the German astronomer J. E. Bode drew attention to a simple numerical relationship concerning the distances of the planets from the sun. The relationship is obtained by writing down the numbers 0, 3, 6, 12, 24, 48, 96 and 192, each of which, apart from the first, is double its predecessor. Next 4 is added to each term giving: 4, 7, 10, 16, 28, 52, 100 and 196. Then taking the earth's distance from the sun as 10 these numbers are approximately proportional to the mean distance of the planets from the sun.

Planets	*According to Bode's Law*	*Actual*
Mercury	4	3·9
Venus	7	7·2
Earth	10	10·0
Mars	16	15·2
Asteroids	28	27·7
Jupiter	52	52·0
Saturn	100	95·4
Uranus	196	191·8

Neptune and Pluto deviate from the predicted Bode distances and, so far, no theoretical basis for this law has been found.

BRACHIATION. Progression by swinging from one hand hold to another.

Callithricidae. One of the two New World monkey familes.

CALORIE. The amount of heat necessary to raise the temperature of one gramme of water by one degree centigrade.

CARBOHYDRATE. Compound of carbon, hydrogen and oxygen, with the latter two in a ratio of two to one; examples are glucose $C_6 H_{12} O_6$, starches and cellulose.

CARNIVORE. A flesh-eating animal.

CATALASE. Enzyme which breaks down the poisonous substance, hydrogen peroxide, formed during plant and animal metabolism to water and oxygen.

CATALYST. A substance that accelerates a chemical reaction but does not become part of the end product.

Cebidae. One of the two New World monkey families.

Ceboidea. Super-family in the sub-order *Anthropoidea*. It includes the New World monkeys of Central and South America; characterized by a flat space between the nostrils.

CELL. The fundamental unit of most living organisms, usually consisting of a nucleus surrounded by cytoplasm and a containing membrane (cell wall).

Cercopithecidae. The family of Old World monkeys in the super-family *Cercopithecoidea*.

Cercopithecoidea. A super-family of primates in the Old World. They differ from New World monkeys in having the nostrils pointing downward and are therefore referred to as catarrhines (downward-nosed).

CEREBRAL CORTEX. Layer of grey matter investing the cerebral hemispheres; well developed only in mammalian brains.

CHLOROPHYLL. One of a group of green pigments found in green plants, a few protozoa and certain bacteria.

CHROMATID. One of the two strands of a chromosome into which a chromosome appears to divide longitudinally during cell division.

CHROMOSOMES. Thread-like bodies containing DNA and protein. They are found in all cells with nuclei and are confined within the latter. They carry the hereditary units of the cell.

COLLOID. Permanent suspension of finely divided particles. Particles usually defined as colloidal range in size from 0·0000001 cm. to 0·0001 cm.

COMBE-CAPELLE TYPE. Human fossils discovered in various parts of Western Europe belonging to the species *Homo sapiens*.

COSMIC RAYS. High speed particles on the atomic scale reaching the earth from outer space mostly from outside the solar system.

CRO-MAGNON TYPE. Fossil men of modern appearance belonging to the species *Homo sapiens*, dating back 35 to 40 thousand years before the present.

CROSSING OVER. A process in which a pair of essentially similar chromosomes break and exchange corresponding parts.

CULTURE. The sum total of habits, customs, beliefs, language and techniques used by man in all fields of activity.

CYTOPLASM. The contents of a cell excluding the nucleus. It is usually a transparent, slightly viscous fluid with inclusions of various sizes.

DENTAL FORMULA. System of numbers used to indicate the numbers and kinds of teeth present in mammals.

DENTITION. Number, kinds and arrangement of teeth in animals and man.

DIASTEMA. Toothless space between two different kinds of teeth, especially the space between incisor and first premolar.

DNA (deoxyribonucleic acid). Found mainly in the nucleus, it is the genetic material of most organisms.

DOMINANT CHARACTER. One of a pair of alternate characters determined by paired genes located on opposite members of a chromosome pair; it exerts its effects so as to exclude manifestations of the alternative (recessive) character.

DOPPLER EFFECT. The apparent change in wavelengths of light (or sound) caused by the motion of the source or of the observer. See RED SHIFT.

Dryopithecus. Genus of fossil anthropoids found in Europe and Asia, probably ancestral to the living apes.

ECOLOGY. Study of the interrelations between animals and plants and their animate and inanimate surroundings.

ELECTROMAGNETIC SPECTRUM. The full range of electromagnetic radiation consisting of radio waves, infra-red rays, visible light and very short wave radiations such as X-rays, gamma-rays and others.

ELECTRON. An elementary particle which is the negatively charged constituent of ordinary matter. The electron is the lightest known particle which possesses an electric charge.

ELEMENT. A substance that cannot be altered by a chemical change so that it is split into any simpler chemical substance. For example, gold, carbon, oxygen and hydrogen are elements.

ENDOPLASMIC RETICULUM. Fine channels bounded by membranes which are dispersed and interconnected throughout the cytoplasm of cells. These structures are associated with the synthesis of proteins.

ENZYME. An organic protein catalyst which enormously speeds up the rate of a metabolic chemical reaction.

EOCENE EPOCH. One of the five geological sub-divisions of the Tertiary period between the Oligocene and Palaeocene; lasting about 22 million years.

ETHOLOGY. The study of behaviour of an animal in its normal environment.

FERMENTATION. Decomposition of organic substances by organisms, especially bacteria and yeasts. For example, decomposition of sugar forming ethyl alcohol and carbon dioxide by yeast in wine-making.

GALAXIES. Star systems each made up of millions of stars, together with gas and dust.

GAMETE (germ cell). A mature male or female reproductive cell whose nucleus and cytoplasm fuses with that of another gamete (constituting fertilization), the resulting cell (zygote) developing into a new individual.

GEIGER COUNTER. An instrument used to detect charged particles and high energy radiation.

GENE FLOW. Spreading of particular genes within a population or between two adjacent populations.

GENES. Units of inherited material situated in the chromosomes.

GENETICS. Study of heredity and variations; of the resemblances and differences between organisms.

GENETIC ISOLATION. Reproductive incompatibility of two or more populations owing to differences in chromosome and gene make-up.

GENOTYPE. The hereditary constitution of an organism as contrasted with the characteristics manifested by the organism.

GENUS. A classification applied to a group of closely related species. Similar genera are grouped in a family.

HAEMOGLOBIN. The red pigment in blood cells able to form a loose temporary combination with oxygen during transport of the latter.

HERBIVORE. A plant-eating animal.

HEREDITY. The transmission from one generation to the next of genetic characteristics causing offspring to resemble parents.

HETEROZYGOUS. Cells which have two different forms of a particular gene in a matched pair of chromosomes. Also used to designate an individual derived from cells that contain unlike genes for the same character.

Hominidae. One of the two families in the primate super-family *Hominoidea*. It includes *Homo sapiens*, the only living species of man and his ancestors (fossil men).

Hominoidea. The super-family of apes and man.

Homo erectus. An extinct species of man which included Java Man, Peking Man and others.

Homo sapiens. The only existing species of man.

HOMOZYGOUS. Cells which have the same form of a particular gene in each of a matched pair of chromosomes. Also an individual derived from cells that contain identical genes for the same character.

HYDROCARBON. Any chemical compound composed only of hydrogen and carbon.

Hylobates. The genus which includes all living gibbons.

INDUSTRY. In relation to prehistoric man, it refers to any set of artifacts which are the work of a single human group.

INFRA-RED RADIATION. Electromagnetic radiation beyond the visible red end of the spectrum: approximately 7,500 to 4 million angstrom units. (1 angstrom = one ten millionth of a millimetre.)

INFUSORIA. A term formerly applied to microscopic organisms found in infusions of organic substances, including various protozoa.

INVERTEBRATES (animals without backbones). A collective term for all animals which are not members of the vertebrata, e.g. amoeba, sponges, jelly-fish, worms, snails, flies, starfish and sea-squirts.

IONOSPHERE. A region of the earth's atmosphere above the stratosphere where the temperature increases. It extends between about 40 and 500 miles above the ground.

ISOTOPES. Any one of two or more kinds of atoms of a chemical element which differ from one another in the number of neutrons in their atomic nuclei. Because of this they have different atomic weights.

LIGHT YEAR. The distance travelled by light in one year. Light moves through space at 186,000 miles per second. A light year is approximately equal to 5,880,000 million miles.

MEIOSIS. The two nuclear divisions by which the number of chromosomes is reduced to one half during the formation of gametes (egg and sperm).

MELANIN. A dark-brown pigment present in many animals and man which in different concentrations gives brown and yellow colouration.

MESOZOIC ERA. The age of the reptiles lasting from about 230 million years to 65 million years before the present.

METABOLISM. In general the chemical processes occurring within an organism or within part of one.

METEORITE. A relatively large extraterrestial body which survives the complete drop to earth without being destroyed.

MILKY WAY. The luminous band stretching across the night sky made up of a large number of stars.

MIOCENE EPOCH. One of the five geological sub-divisions of the Tertiary period between the Pliocene and Oligocene; lasting about 12 million years.

MITOSIS. The usual process by which a nucleus divides into two. Each chromosome duplicates, probably before the beginning of the mitosis, and mitosis involves separation of resulting duplicates so that one goes into each daughter nucleus. As a result the two daughter nuclei have an identical complement of chromosomes and hence of genes.

MOLECULE. The smallest part of an element or compound that can exist as a free and separate substance, yet retain all the chemical properties of the substance.

MULTIPAROUS. Bearing more than one child.

MUTATION. New, genetic feature in an individual produced by changes in the individual's genes or chromosomes. Only mutations in the gametes can produce heritable changes.

NATURAL SELECTION. The principal mechanism of evolutionary change suggested by Charles Darwin in 1859. It is a complex of processes by which the collective factors of the environment eliminate those individuals least fitted to that environment.

NEANDERTHAL MAN. A prehistoric sub-species of the species *Homo sapiens* (about 130,000–40,000 B.P.)

NEBULA. A mass of tenuous gas in space, together with what is loosely termed 'dust'.

NEOLITHIC. Post-glacial phase of human history which began about ten thousand years ago and marked by the beginning of agriculture.

NUCLEIC ACIDS. Compounds of pentose sugar, phosphoric acid and nitrogen-containing bases (purines and pyrimidines) of high molecular weight, characteristic of all living things.

NUCLEOLUS. Small dense body, one or more of which occurs inside the resting nucleus, containing ribose nucleo-protein.

NUCLEOTIDE. A molecule containing a nitrogenous base, a pentose sugar and a phosphate; e.g. adenine nucleotide consists of adenine (base), deoxyribose (sugar) and a phosphate.

NUCLEUS. Body containing the chromosomes. Present in nearly all cells of plants and animals and probably in bacteria; not in viruses.

OCCIPITAL BONE. Bone forming the back part of the head of vertebrates.

OLIGOCENE EPOCH. One of the five geological sub-divisions of the Tertiary period, between the Miocene and Eocenes lasting about 11 million years.

OMNIVOROUS. Eating a diet both of plant, and animals.

OREOPITHECUS. A genus of fossil primates belonging to the super-family of man and apes (*Hominoidea*) from the Pliocene epoch; not clearly related to either the *Hominidae* or *Pongidae*.

OVUM. An egg; the female sex cell.

PALAEOCENE EPOCH. Earliest of the five sub-divisions of the Tertiary period; lasting about 7 million years.

PALAEOLITHIC. The Old Stone Age of man beginning from about 1½ million years ago and ending with Neolithic times (the beginning of agriculture).

PALAEONTOLOGY. The study of fossil animals and plants and their distribution in time.

PARIETAL BONE. One of a pair of bones on the roof of a vertebrate skull; in man the two parietals form much of the top and sides of the cranium.

PARTURITION. Bringing forth young.

Pelycosauria. An order of primitive mammal-like reptiles now extinct and known from rocks of the Upper Carboniferous and Lower and Middle Permian periods.

PEPTIDE. Compound formed from two or more amino acids linked by peptide bonds. The term is usually reserved for molecules of low molecular weight.

PEPTIDE BOND. The bond (—CO—NH—) formed when the carboxyl group (COOH) of one amino acid reacts with the amino group (NH_2 of another with the elimination of a molecule of water.

PHENOTYPE. The characteristics manifested by an organism as contrasted with the set of genes possessed by it.

PHOTOSPHERE. The visible surface of the sun. Its temperature is about 6,000 degrees centigrade.

PHOTOSYNTHESIS. Synthesis by green plants of organic compounds from water and carbon dioxide using energy absorbed from sunlight.

PHYSIOLOGY. The study of the processes which go on in living organisms.

Pithecanthropus erectus. Java Man member of the species *Homo erectus*.

PLACENTATION. Manner of formation, arrangement and structure of the placenta ('after-birth') in mammals.

PLEISTOCENE EPOCH. A recent geological epoch lasting about 3 million years and ending 15,000 to 20,000 years ago; a sub-division of the Quaternary period.

PLIOCENE EPOCH. The last of the five geological sub-divisions of the Tertiary period, lasting about 10 million years.

Pondigae. A family of the order primates which includes all living apes and fossil apes.

Posimii. A sub-order of the order primates which includes the tree shrews, lemurs, lorises and tarsiers.

PRIMATES. The order of mammals which includes tree shrews, lemurs, tarsiers, monkeys, apes and man.

Proconsul. An important genus of Miocene apes.

PROTEIN. Very complex organic compound, composed of numerous amino acids. Proteins of many different kinds are present in all living things, making up a considerable proportion of their dry weight. A protein molecule may be made up of hundreds or thousands of amino acid molecules which are usually of about twenty different kinds.

PROTEINOID. Compounds of high molecular weight produced by Dr S. W. Fox by heating mixtures of amino acids containing high proportions of aspartic and glutamic acids.

PROTON. An elementary particle which is the positively charged constituent of ordinary matter. Together with the neutron, the proton is a building-stone of all atomic nuclei.

QUADRUPED. Any four-footed animal.

QUASAR. A very distant highly luminous object emitting strong radio waves.

RADIO-CARBON DATING. A method of determining the age of biological material by establishing the relative proportion of carbon with the normal atomic weight of 12 to that of radioactive carbon having an atomic weight of 14.

RADIO-ISOTOPE. A radioactive isotope as distinguished from a stable isotope of an element. Atomic nuclei are of two types, unstable and stable. Those in the former category are said to be radioactive and eventually are transformed by radioactive decay into the latter.

RADIO TELESCOPE. An instrument used for studying the radio waves from space.

RED SHIFT. When a luminous body is moving away its spectral lines will be shifted over to the red or long wave end of the spectrum. The greater the shift, the greater the speed at which the body is moving away. See DOPPLER EFFECT.

REDUCTION DIVISION. The two nuclear divisions leading to the formation of the gametes by which the number of chromosomes in each gamete is reduced to one half.

RESPIRATION. The use of oxygen by the cell or body with carbon dioxide and water as chief end-products.

REVERSING LAYER. The gaseous layer above the bright surface (photosphere) of the sun.

RIBOSOMES. Fine granules composed of protein and RNA found either attached to the endoplasmic reticulum or free in the cytoplasm of cells. They are associated with the synthesis of proteins in the cell.

RNA (ribonucleic acid). Found both in the nucleus and cytoplasm of cells, it occurs in three forms each of which plays a role in the synthesis of proteins. In some viruses RNA is the genetic material but no case is known in which a compound other than a nucleic acid forms the genetic material.

SAVANNA. Open grassland with few trees characterized by low rainfall in tropical and subtropical regions.

SEX LINKAGE. Genes that determine characters other than sex which are permanently associated with the sex chromosones.

SICKLE-CELL ANAEMIA. An hereditary disease principally found in countries where malaria is endemic, characterized by abnormal blood cells.

SIDEREAL DAY. The earth's rotation period as measured with reference to the stars. The sidereal day has 23 hours 56 minutes.

Sinanthropus pekinensis. Peking Man member of the species *Homo erectus*.

SOLAR CONSTANT. A unit of energy for measuring the amount of solar radiation received on the earth's surface. It equals 1·94 calories per minute per square centimetre.

SOLO MAN. A fossil man of the Upper Pleistocene known from skull fragments found in Indonesia.

SOLAR SYSTEM. The system made up of the sun, the planets, satellites, comets, meteoric bodies and interplanetary dust and gas.

SOLAR WIND. A stream of atomic particles of solar origin that is caused by eruptions on the sun's surface.

SPECIES. A group of organisms which interbreeds and which is reproductively isolated from other groups.

SPECTROSCOPE. An instrument incorporating a system of glass prisms or a ruled grating that resolves light into its spectrum colours.

SPERM. The male sex cell.

STRATOSPHERE. A region of temperature increase above the tropopause. It extends to a height of about forty miles.

SUPERNOVA. A star which appears to disintegrate, sending much of its material into space. Supernovae show a rapid increase of light, sometimes becoming a thousand million or more times brighter than the sun.

TERRESTRIAL. Normally applied to animals spending most of their lives on the ground.

TERTIARY PERIOD. A geological period lasting about 62 million years; divided into Pliocene, Miocene, Oligocene, Eocene and Palaeocene epochs.

TETRAPODS. All those vertebrates which normally have four limbs, namely amphibians, reptiles, birds and mammals.

THERMONUCLEAR REACTION. A nuclear fusion reaction which occurs between various nuclei of the light elements when they are constituents of a gas at a very high temperature. Thermonuclear reactions, the source of energy generation in the sun and the stable stars, are utilized in the fusion bomb.

TROPOPAUSE. A region of nearly constant temperature immediately above the troposphere.

TROPOSPHERE. The part of the atmosphere in which we live, and which is the theatre for weather as we know it. It extends to a height of about seven miles.

VAN ALLEN RADIATION ZONES. Zones around the earth in which electrically charged particles are trapped and accelerated by the earth's magnetic field.

VILLAFRANCHIAN. The earliest part of the Pleistocene before the First Glaciation, associated with an important assemblage of animals.

WÜRM. The fourth and final glacial phase of the Pleistocene, named after an Alpine location.

ZYGOTE. The fertilized ovum, before it undergoes further differentiation.

Bibliography and Allied Reading

CHAPTER 1

DICKE, R. H. 'Gravitation and the Universe'. *Science Journal*, October 1966.

DONALD, E. L., VIELE, D. and ELDRENKAMP, L. B. 'The Lunar Orbiter Mission to the Moon'. *Scientific American*, May 1968.

DUNBAR, C. O. *The Earth*. Weidenfeld and Nicolson, 1966.

EGGEN, O. J. 'Stars in Contact'. *Scientific American*, June 1968.

ESHLEMAN, R. VON. 'The Atmospheres of Mars and Venus'. *Scientific American*, March 1969.

GOLDBERG, L. 'Ultraviolet Astronomy'. *Scientific American*, June 1969.

GREENBERG, M. J. 'Interstellar Grains'. *Scientific American*, October 1967.

HEWISH, A. 'Pulsars'. *Scientific American*, October 1968.

HONG-YEE CHIU. 'The Evolution of the Universe'. *Science Journal*, August 1968.

KING-HELE, D. 'The Shape of the Earth'. *Scientific American*, October 1967.

LOVELL, SIR BERNARD. *Our Present Knowledge of the Universe*. Manchester University Press, 1967.

LOVELL, SIR BERNARD. 'The Moon and After'. *Science Journal*, May 1969.

LYTTLETON, R. A. 'The Origin of the Moon'. *Science Journal*, May 1969.

MCVITTIE, G. C. *Facts and Theory in Cosmology*. Eyre & Spottiswoode, 1961.

MOORE, P. *The Amateur Astronomer's Glossary*. W. W. Norton & Company Inc., 1967.

MOORE, P. *Astronomy*. Oldbourne Book Co. Ltd., 1967.

MOORE, P. *Basic Astronomy*. Oliver & Boyd, 1967.

MURCHIE, G. *Music of the Spheres*. Dover Publications, 1967.

PEEBLES, P. J. E. and WILKINSON, D. T. 'The Primeval Fire Ball'. *Scientific American*, June 1967.

REDDISH, V. C. 'Infant Stars'. *Science Journal*, July 1968.

RONAN, C. A. *Invisible Astronomy*. Eyre & Spottiswoode, 1969.

SMITH, G. F. 'Pulsars'. *Science Journal*, June 1969.
THORNE, K. S. 'Gravitational Collapse'. *Scientific American*, November 1967.
VIORST, J. *The Changing Earth*. Bantam Books, 1967.
WEYMANN, J. R. 'Seyfert Galaxies'. *Scientific American*, January 1969.

CHAPTER 2

ALLFREY, V. G. and MIRSKY, A. E. 'How Cells make Molecules'. *Scientific American*, September 1961.
ARISTOTLE. As quoted from *The Generation of Animals* in A. I. Oparin, *The Origin of Life on Earth*, New York, Academic Press, 1957.
ARNON, D. I. 'Photosynthesis as an Energy Conversion Process'. *Frontiers of Modern Biology*, Houghton Mifflin Co., 1962.
ASIMOV, I. *The Genetic Code*. The New American Library, 1963.
BARRY, J. M. *Molecular Biology and the Chemical Control of the Living Cell*. Prentice-Hall Inc., 1964.
BERNAL, J. D. *The Origin of Life*. Weidenfeld and Nicolson, 1967.
CALVIN, M. 'Life's Origin and its Implication'. *Science Journal*, July 1968.
CLARK, B. F. C. and MARCKER, K. A. 'How Proteins Start'. *Scientific American*, January 1968.
CRICK, F. H. C. 'On the Genetic Code'. *Science*, 8 February 1963.
DUCHESNE, J. 'Meteorites and Extraterrestrial Life'. *Science Journal*, April 1969.
EGLINTON, G. and CALVIN, M. 'Chemical Fossils'. *Scientific American*, January 1967.
FOX, S. W. *A Chemical Theory of Spontaneous Generation. Aspects of the Origin of Life*. Pergamon Press, 1960.
GOULIAN, M. 'Synthesis of Viral DNA'. *Science Journal*, March 1969.
HANAWALT, C. P. and HAYNES, H. R. 'The Repair of DNA'. *Scientific American*, February 1967.
HOOKE, R. 'Of the Schematisme or Texture of Cork, and of the Cells and Pores of Some Other Such Frothy Bodies', as quoted in *Great Experiments in Biology*, Prentice-Hall, Inc., 1955.
HOTCHKISS, R. D. and WEISS, E. 'Transformed Bacteria'. *Scientific American*, November 1956.
KEOSIAN, J. *The Origin of Life*. Chapman & Hall, 1964.
KORNBERG, A. 'The Synthesis of DNA'. *Scientific American*, October 1968.
LANDSBERG, H. E. 'The Origin of the Atmosphere'. *Scientific American*, August 1953.
MERRIFIELD, R. B. 'The Automatic Synthesis of Proteins'. *Scientific American*, March 1968.
MILLER, S. L. 'Production of Amino Acids under Possible Earth Conditions'. *Science*, Vol. 117, 15 May 1953.
OPARIN, A. I. *Origin of Life*. Dover Publications, 1953.
PASTEUR, L. *Great Experiments in Biology*, ed. M. Gabriel and S. Fogel. Prentice-Hall Inc., 1955.

PFEIFFER, J. E. 'Enzymes'. *Scientific American*, December 1958.
REDI, F. *Experiments on the Generation of Insects*. Open Court Publishing Co., 1909.
RHODES, F. H. T. *The Evolution of Life*. Pelican Books, 1965.
SAGAN, C. 'Simulating Extraterrestrial Environments'. *Science Journal*, March 1968.
SWANSON, C. P. *The Cell*. Prentice-Hall Inc. (Foundations of Modern Biology Series), 1964.
WALD, G. 'Life and Light'. *Scientific American*, October 1959.
WATSON, J. D. *The Double Helix*. Weidenfeld and Nicolson, 1968.
WATSON, J. D. *Molecular Biology of the Gene*. W. A. Benjamin Inc., 1965.
WATSON, J. D. and CRICK, F. J. C. 'Molecular Structure of Nucleic Acids'. *Nature*, Vol. 171, 1953.
YANOFSKY, C. 'Gene Structure and Protein Structure'. *Scientific American*, May 1967.

CHAPTER 3

ARISTOTLE, *Histories about Animals. The Basic Works of Aristotle*, Richard McKeon. Random House, 1941.
AUERBACH, C. *Genetics in the Atomic Age*. Oliver & Boyd, 1965.
CROW, J. F. 'Ionizing Radiation and Evolution'. *Scientific American*, September 1959.
DARWIN, C. *The Voyage of the Beagle*. New York, Harper and Brothers, 1959.
DARWIN, C. *The Autobiography of Charles Darwin*. New York, Dover Publications, 1958.
DARWIN, C. *The Origin of Species*. New York, Mentor Books, 1958.
ILTIS, H. *The Life of Mendel*. New York. W. W. Norton Company, Inc., 1932.
LACK, D. 'Darwin's Finches'. *Scientific American*, April 1953.
LAMARCK, J. B. as quoted in *The Major Achievements of Science* by A. E. E. McKenzie. Cambridge University Press, 1960.
LASKER, G. W. *Human Evolution*. Holt, Rinehart and Winston Inc., 1963.
MAZIA, D. 'How Cells Divide'. *Scientific American*, September 1960.
MENDEL, G. 'Experiments in Plant Hybridisation'. *Classic Papers in Genetics*, New York, Prentice-Hall Inc., 1959.
MIRSKY, A. E. 'The Discovery of DNA'. *Scientific American*, June 1968.
MULLER, H. J. 'Radiation and Human Mutation'. *Scientific American*, November 1955.
OLSON, E. C. *The Evolution of Life*. The New English Library Ltd, 1966.
RYAN, F. J. 'Evolution Observed'. *Scientific American*, October 1953.
SAVAGE, J. M. *Evolution*. University of Southern California, 1963.
SEILACH, A. 'Fossil Behavior'. *Scientific American*, August 1967.
SIMPSON, G. G. *The Meaning of Evolution*. The New American Library, 1951.
SOLBRIG, O. T. *Evolution and Systematics*. Macmillan Co. Inc., 1966.
TAYLOR, J. H. 'The Duplication of Chromosomes'. *Scientific American*, June 1958.

WALLACE, B. *Chromosomes, Giant Molecules and Evolution.* Macmillan Co. Inc., 1967.

WOOD, B. and EDGAR, S. R. 'Building a Bacterial Virus'. *Scientific American,* July 1967.

CHAPTER 4

BARNETT, A. *The Human Species.* Pelican Books, 1965.

BINFORD, S. R. and BINFORD, L. R. 'Stone Tools and Human Behavior'. *Scientific American,* April 1969.

BRACE, C. L. and MONTAGU, M. F. A. *Man's Evolution.* Macmillan, 1965.

BRAIDWOOD, R. J. 'The Agricultural Revolution'. *Scientific American,* September 1960.

BUETTNER-JANUSCH, *The Origin of Man.* Wiley, 1965.

CAMPBELL, B. *Human Evolution.* Heinemann Educational Books, 1967.

CLARK, J. 'Early Man in Africa'. *Scientific American,* July 1958.

CLARK, G. and PIGGOTT, S. *Prehistoric Societies.* Hutchinson, 1967.

CORNWALL, I. W. *Prehistoric Animals and Their Hunters.* Faber & Faber, 1968.

CROOK, J. 'Evolutionary Change in Primate Societies'. *Science Journal,* June 1967.

EDWARDS, W. N. *The Early History of Palaeontology.* British Museum Natural History.

HOWELLS, W. W. *Mankind in the Making.* Pelican Books, 1967.

HOWELLS, W. W. '*Homo Erectus*'. *Scientific American,* November 1966.

HOWELLS, W. W. *Ideas on Human Evolution.* Harvard University Press, 1961.

KIELAN-JAWOROWSKA, Z. 'Fossils from the Gobi Desert'. *Science Journal,* July 1969.

LANNING, P. E. and PATTERSON, C. T. 'Early Man in South America'. *Scientific American,* November 1967.

LEAKEY, L. S. B. *Adam's Ancestors.* Methuen, 1960.

LEROI-GOURHAN, A. 'The Evolution of Palaeolithic Art'. *Scientific American,* February 1968.

NAPIER, J. 'The Antiquity of Human Walking'. *Scientific American,* 1967.

SIMONS, E. L. 'The Early Relatives of Man'. *Scientific American,* July 1964.

SIMSON, E. L. 'The Earliest Apes'. *Scientific American,* December 1967.

OAKLEY, K. *Frameworks for Dating Fossil Man.* Weidenfeld & Nicolson, 1969 (3rd Ed.).

WASHBURN, S. L. 'Tools and Human Evolution'. *Scientific American,* September 1960.

WASHBURN, S. L. *Social Life of Early Man.* Aldine Publishing Company, 1961.

WHEAT, B. J. 'A Paleo-Indian Bison Kill'. *Scientific American,* January 1967.

CHAPTER 5

ACKERKNECHT, E. H. 'White Indians'. *Bulletin of the History of Medicine,* XV, 1944.

ALLEE, W. C. *The Social Life of Animals.* Beacon Press, 1958.

CAVALLI-SFORZA, L. L. '"Genetic Drift" in an Italian Population'. *Scientific American*, August 1969.

COLE, S. *Races of Man*. British Museum Natural History.

COON, C. S. *The Living Races of Man*. Jonathan Cape, 1966.

DOBZHANSKY, T. *Mankind Evolving*. Yale University Press, 1965.

DOBZHANSKY, T. *Heredity and the Nature of Man*. Harcourt, Brace and World, 1964.

DEVORE, I. *Primate Behaviour*. Holt, Rinehart and Winston, 1965.

FOX, M. *The Personality of Animals*. Penguin Books, 1952.

MEAD, M. *An Anthropologist at Work*. Atherton Press Paper Back, 1959/1966.

MEAD, M. *Continuities in Cultural Evolution*. Yale University Press, 1965.

MEAD, M., DOBZHANSKY, T., TOBACH, E., LIGHT, R. E. *Science and the Concept of Race*, Columbia University Press, 1968.

MONTAGU, M. F. A. *Man's Most Dangerous Myth: The Fallacy of Race*. Meridian Books, 1965.

MONTAGU, M. F. A. *The Humanization of Man*. The World Publishing Company, 1962.

MONTAGU, M. F. A. (editor). *The Concept of Race*. Collier-Macmillan Ltd., 1965.

MONTAGU, M. F. A. *Race, Science and Humanity*. Van Nostrand, 1963.

MONTAGU, M. F. A. *Culture and the Evolution of Man*. New York, Oxford University Press, 1962.

SAHLINS, D. M. 'The Origin of Society'. *Scientific American*, September 1960.

VAN LAWICK-GOODALL, J. 'The Behaviour of Free-Living Chimpanzees in the Gombe Stream Reserve'. *Animal Behaviour Monographs*, Part 3, Vol. 1. 1969.

WASHBURN, S. L. and DEVORE, I. 'The Social Life of Baboons'. *Scientific American*, June 1961.

CHAPTER 6

ALLAND, A. *Evolution and Human Behaviour*. Tavistock Publications, 1969.

DOBZHANSKY, T. *Evolution, Genetics and Man*. John Wiley & Sons, Inc., 1955.

DOBZHANSKY, T. *Heredity and the Nature of Man*. George Allen & Unwin Ltd, 1965.

WALLACE, B., DOBZHANSKY, T. *Radiation, Genes and Man*. Henry Holt & Co., Inc., 1959.

Index

Figure numbers are given in italic numerals